Fly Fishing
The West's Best
Trophy Lakes

A Fly Fisher's Comprehensive Guide
To 50 Of The Best
Trophy Lakes and Reservoirs

By
DENNY RICKARDS

•

Edited by Pat Hoglund

FLY FISHING THE WEST'S
BEST TROPHY LAKES

A Fly Fisher's Comprehensive Guide
To 50 Of The Best Trophy Lakes and Reservoirs

By
DENNY RICKARDS
•
Edited by Pat Hoglund
Associate Editor - Frank Phillips
Technical Consultant - David Nolte
•
All Uncredited Photos By Denny Rickards
•
A Stillwater Productions Publication
P.O. Box 470
Fort Klamath, OR 97626

Copyright 1999 • Stillwater Productions • Denny Rickards
Printed in Hong Kong
Book Design: Nancy L. Doerrfeld-Smith
Front Cover: Denny Rickards with a six-pound rainbow
from a high desert western lake.

Library of Congress Cataloging-In-Publication Date
ISBN 0-9656458-1-9

Dedication

To Gail:
Your love and thoughtful consideration
while I tramped around the country
kept me inspired
and made this book possible.

Acknowledgments

To Dave Freel,
my partner in Stillwater Productions,
for all the times we spent on the road traveling
to every lake, reservoir, pond and watering hole that
might have contained a big trout, for the laughs,
encouragement and gourmet meals we shared
while researching this book.

To Dave Nolte, for sharing ideas for the book
and for all your help with our photo sessions
on the water.

And especially to my wife Gail,
for being there for me.

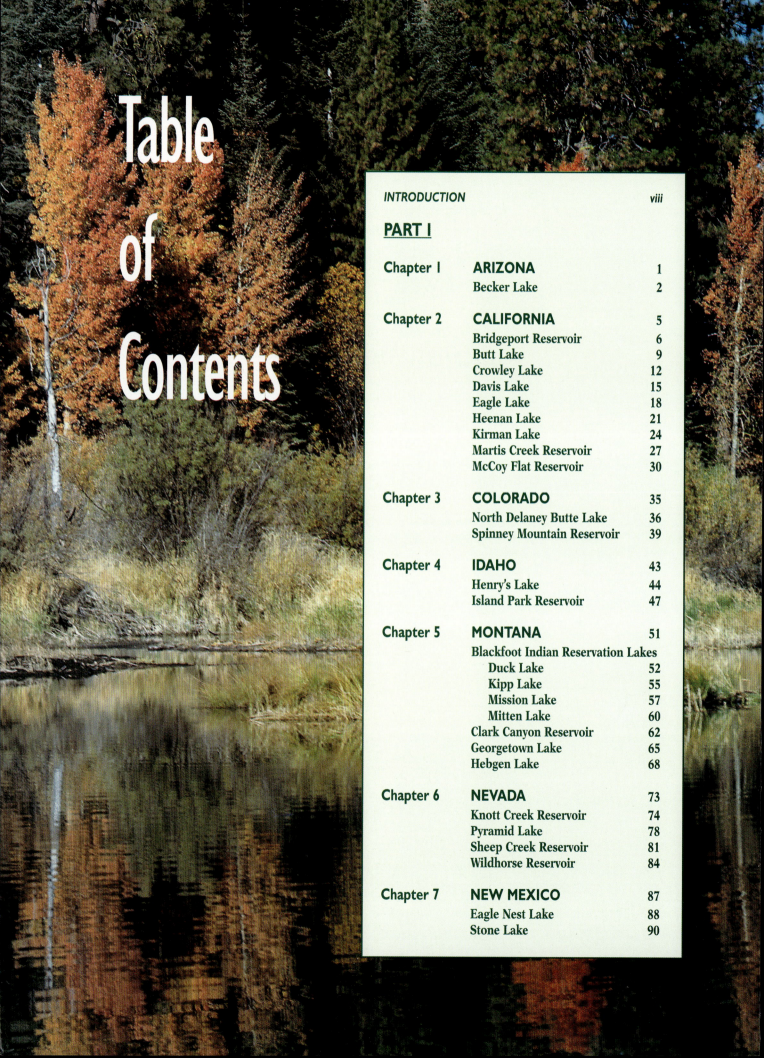

Table of Contents

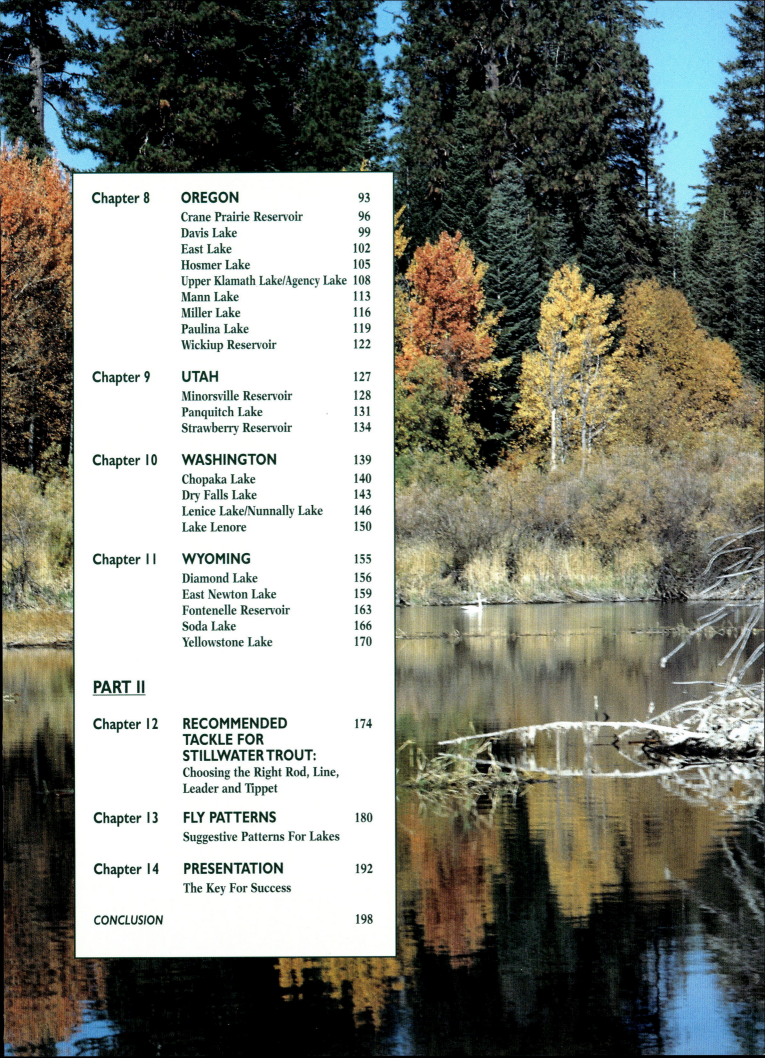

Introduction

To catch large, bragging-sized trout is a challenging task in itself. It is also a life-long passion of mine. For the better part of two decades I have spent countless hours pursuing big trout, and through the course of time, these creatures have taught me many lessons. But, one obvious fact holds true: your odds improve dramatically if your efforts are applied to those lakes and reservoirs where large numbers of big trout live. It's a simple, but often overlooked lesson.

In my first book, "Fly Fishing Stillwaters for Trophy Trout", the focus was to help stillwater fly fishermen become successful as predators of large trout, those that have avoided anglers over the years. In this book, "Fly Fishing The West's Best Trophy Lakes", you will find not only a comprehensive guide to the lakes, but the information necessary to be successful once you're there. I've also included recommended tackle, fly patterns, trout locations, lake chemistry, depths, habitat, insects, species of trout, and average sizes, along with seasonal time frames for the best fishing.

You hold in your hands a guide that will improve your odds of catching some of the largest trout you will ever encounter.

In the Western United States there are hundreds of quality lakes and reservoirs supporting big trout and intense numbers of aquatic insects, yet no two fisheries will be the same. Each is challenging, productive, yet different. Intimidating as each lake, reservoir and pond may be, they always seem to differ from one another as much as ripples in a stream, always changing, rarely predictable and mostly a mystery. These are but a few of the characteristics we must face, yet each helps to build that anticipation that draws us to the sport.

All lakes, whether natural bodies of water or man-made reservoirs, are forever undergoing changes. That is normal because Mother Nature's forces are constantly at work reflecting her moods such as droughts, flooding, winter and summer fish kills, along with radical changes in water temperature, all of which affect the quality of fishing. Many go in cycles reflecting a particular management plan, increased fishing pressure and so on. Most lakes are somewhat temperamental to begin with and are often inconsistent with up and down cycles. They produce well for a period of time then tend to have some off years where fishing is only so-so. Many of the lakes I have fished were subject to man's tinkerings and irresponsible acts while others have gone through eradication projects to eliminate rough or non-native species.

Qualifying fifty of the best is a near-impossible task, one that invites controversy and is certainly a topic for debate. If I asked one hundred fly fishermen to identify their most productive lakes, no two lists would be the same. I think that is something we can all agree on.

Over the past fifteen years, I have fished, explored and challenged trout in over 400 lakes, reservoirs and ponds in the Western United States. To justify a mention in this book, certain criteria had to be met. Each lake had to be capable of supporting big trout, rich enough in aquatic insects or other food sources to sustain their numbers and most importantly, for anglers to successfully imitate with flies. From this list, fifty of the best stillwater fisheries surfaced. Each was ranked depending on the overall quality of the fly fishing. Many lakes were capable of growing big trout, but didn't fly fish well. High altitude, acidic type lakes as well as deep bodied waters seldom qualified. Most of the lakes and reservoirs that met the criteria were high desert lakes containing alkaline waters rich in nutrients. Most were shallow with weedy shoreline areas, intense aquatic insect hatches and often displayed algae blooms common with warmer summer temperatures. Healthy numbers of fast growing trout were almost always present as were a large supporting cast of big trout.

Rightly so, some fly fishermen I spoke with were concerned their favorite water would forever be exposed while others wondered why I left their most productive lake off the list. I can assure you, much thought went into compiling this list and if undue angling pressure would adversely affect it, it was not included. To that end, many lakes holding some exceptional trout that were referred to me in confidence along with those I have discovered on my own will remain a secret because I'm convinced angling pressure would jeopardize their fragile existence. Fee lakes were not considered and only those open to public fishing were included.

As I travel and explore new waters, this list will obviously need changing, but I'll leave that as a project for another day. Please keep in mind the lakes and reservoirs listed in this book are not my list of the *best* fifty lakes in the West, but fifty *of* the best.

— Denny Rickards

Arizona—Big trout are only part of the promise Arizona has to offer.

Chapter 1

Arizona

The White Mountains Fort Apache Reservation is host to most of the better stillwater opportunities for fly fishermen in Arizona. Lakes such as Crescent, Big, Sunrise, Tonto, Cienega and Hawley are only a few of the reservation lakes that are excellent fly fishing waters. Many, however, don't harbor enough big trout to qualify for trophy status.

But in fairness to the state of Arizona there is one lake that keeps popping up that I'm quite sure would have made the Top 50 if I had the time and opportunity to fish it. The name is Earl Park, adjacent to Hawley Lake on the reservation. It contains Apache, brown, rainbow and brook trout, some of which get quite large. I've been told this is a rich lake with lots of aquatic insects with shallow weedy areas and supports some large trout. You can bet this one is on my agenda for the near future.

If you are a back packing enthusiast, the Mogollon Rim Lakes located in the Apache Sitgreaves National Forest offers breath-taking scenery, solitude and outstanding fly fishing. There are a variety of lakes, but Chevelon Lake is the best. It's approximately a mile hike, but the rainbows and browns that live in this 200-acre lake are said to be worth it. Again, this is also on my list of lakes to fish.

There are obviously some other hidden gems around the state that I haven't discovered. Knowing that Arizona's scenic areas are well worth the time spent exploring, I look forward to testing these waters another day.

BECKER LAKE
Arizona

Dave Freel with a 6 pound rainbow that took a Seal Bugger.

Only a short cast away from the town of Springville, Becker Lake is managed as a trophy lake restricted to flies or lures only. There is a four fish limit and any fish under 14 inches must be released unharmed. Becker is really an irrigation reservoir that covers between 80 and 90 acres of water depending on the time of year. It's a relatively shallow lake with an average depth around 9 to 10 feet and supports decent hatches of mayflies, caddisflies and midges with leeches and dragonfly nymphs also available.

Becker's reputation is based on its fish, not its tranquil setting. The lake holds rainbow and brown trout that average 14 to 20 inches, but bigger fish from 5 to 7 pounds are available. The last time I fished this lake was in late October of 1996, and using my Seal Bugger in a variety of colors, I managed to land quite a few fish from 18 to 20 inches including one big brown around 6 pounds.

The lake opens the last weekend in April and closes the end of November. The best fly fishing occurs from opening day and lasts into June or longer if the water remains cool. Summer months are slower, but the bite improves again in late September and will last well into November or until the temperatures drop below freezing.

A check with a local fly shop indicated most anglers prefer perhaps the most popular fly in Arizona, the Peacock Lady, along with Hare's Ears and Zug Bugs in sizes 10 to 14. The trout at Becker can be taken on leech patterns, Woolly Buggers and a variety of large streamers in sizes 6 or 8 as well.

Because of its size, most of Becker can be fished from shore, but a small boat or a float tube puts you over more fish. I find an intermediate line is ideal for this shallow fishery, but conditions may require you to use either the clear transparent or floating line. Leaders from 10 to 12 feet matched with 4x and 5x tippets will cover most situations you will encounter.

A 4 pound 'bow landed on Denny's Stillwater Nymph.

Becker is located just of Highway 180 a few miles north of Springville. Camping is not permitted at the lake, but there is a private campground close by. Lodging and fishing information is available in Springville.

This 5 pound brown was cruising shallow water when the author showed him an olive Seal Bugger.

Throughout the western states it seems most little towns situated in or near trout country harbor small lakes and reservoirs loaded with big fish. Springville is one of those. And if Becker is off for some reason, there are a host of other quality lakes less than an hour's drive from here.

Becker Lake Summary

SEASON: Opens last Saturday in April and closes the end of November.

TROUT: Rainbow and brown trout with fish up to 7 pounds possible. Average fish runs 14 to 20 inches.

LAKE SIZE: Approximately 90 acres when full with the average depth 9 to 10 feet.

RECOMMENDED FLY LINES: Intermediate slow sink, fast sinking type II and III or floating lines are best.

RECOMMENDED FLY PATTERNS: I used my Seal Buggers in size 8, Stillwater and Callibaetis nymphs in size 10, A.P. Emerger size 12, but a host of standard patterns work here.

BEST TIMES: Opening day into early June and again late September into November or until it freezes.

California—The golden state offers anglers a wide assortment of challenges including big trout.

Chapter 2

California

California is a state blessed with an abundance of quality fly fishing lakes many of which are located on the eastern slope of the High Sierras extending from the town of Bishop all the way to the Oregon border. U.S. Highway 395 runs through the eastern portion of the state linking a substantial number of rich, fertile lakes and reservoirs for anglers to explore.

Much of the eastern half of the state is high chaparral, semi-arid country with sagebrush sprinkled across the dessert floor and pine forests scattered along the base of the Sierras. This is also fly fishing country with plenty of shallow, alkaline lakes most of which are food factories providing fast growth resulting in big trout.

Rainbows, browns and cutthroats are the beneficiaries of the habitat that also produce numerous aquatic insect hatches for fly fishermen to imitate.

In the higher elevations, brook and golden trout are the dominant species, and solitude and unparalleled beauty are offered while pursuing these hardy fish.

Some of the Sierra lakes are deep, large bodies of water that support huge trout, some over 20 pounds on occasion. Anglers often fall prey to the notion that big trout are only found in deep lakes with bottoms reaching 200 to 300 feet. Obviously, if you fish Crowley Lake, or Bridgeport Reservoir, you will know that's not true. Many of the larger, deeper lakes have their window of opportunity, but unfortunately don't fly fish well. Conversely, there are countless shallow lakes rich with nutrients and forage that grow incredibly large trout to keep you pacified and coming back for more. Add to that the scenic wonders and points of interest in this state and you're apt to find traveling from lake to lake interesting while pursuing not only trout, but other species that dominate the Golden State.

BRIDGEPORT RESERVOIR
California

Bridgeport Reservoir, an unknown to most stillwater anglers, is a rich and fertile lake with decent numbers of double digit trout.

Located on the eastern slope of the High Sierras behind the little town of Bridgeport, this reservoir is virtually unknown to stillwater fly fishermen. Perhaps, it's because the area is surround by quality fisheries. The East Walker River which enters and drains the reservoir has long been the magnet for fly fishermen. Browns over 20 pounds are landed quite frequently from Twin Lakes just 13 miles north of Bridgeport, and Kirman Lake to the east produces some of the biggest brook trout landed in the state each year. Bridgeport Reservoir, on the other hand, has always been popular with trollers and bait fishermen with double-digit trout fairly common.

The reservoir lies in a shallow barren valley surrounded by sagebrush and sandy flats at an elevation of 6,200 feet. Its trout and their growth rates are classic examples of nutrient-rich lakes with shallow shorelines. It's weedy at the western end of the lake and thick with algae bloom during the summer months. The deepest water is at the dam end and in the old streambed channels that run down the middle. The rest of the lake is fairly shallow averaging about 8 feet in depth when full.

This is a rainbow and brown trout fishery that gets annual plants of catchable size trout each year along with some brood stock averaging 4 to 5 pounds. The browns grow fat here on a diet of chubs and other forage fish with 8 to 10 pound fish not at all uncommon.

Midges are the only aquatic insects I've seen in any numbers with Callibaetis mayflies, damsels, and dragonflies playing a lesser role for fly fishermen to imitate. The preferred fly patterns of those anglers who are regulars here seem to be black Leeches, Woolly Buggers and Matuka-style streamers. Black, brown and olive are the preferred colors of both trout and fly fishermen. I've enjoyed my best action using my Seal Buggers early and again late in the day and switching to my A.P. Emerger in size 10 to 12 during mid-day hours.

You can use a variety of lines here depending on whether you want to probe the depths or work near the surface. Fly fishermen working the dam area for big browns use medium to fast sinking lines while those working the shoreline areas are better off with the intermediate or clear transparent lines. I've always enjoyed decent success working the weed beds off the mouth of the Upper East Walker River where it enters the lake. For this, I use the intermediate line and fish the top 6 feet, with my clear line in reserve if the lake gets flat.

Like all nutrient-rich lakes, you can usually cheat on leader length and tippet size and Bridgeport is no different. I use a 12 foot leader with 3x fluorocarbon tippet, which tests about 8 pounds. It's usually enough to get these fish out when they hang up in the weed beds. As always, adjust your leaders and tippets to the conditions present.

If I had to pick a prime time for Bridgeport it would be late May through June and again late September through October. The summer months are iffy when trout hug the deeper channels in the middle of the lake.

This 3 pound brown was fooled by Denny's Olive A.P. Emerger while feeding just below the surface on midge pupae.

Water temperatures will be coldest in the spring so use slow retrieves with your flies, but remember to adjust your speed relative to the pattern being fished. The angle and depth you fish it should be relative to conditions above and below the surface. I almost always find a long slow or short, semi-jerky pull consistent with the bugger type flies. If I fish the smaller insect patterns, the hand-twist retrieve is better suited.

In normal water years the reservoir is about 3,000 acres. You can wade it, but it is best fished from a small boat or a float tube. Expect windy afternoons regardless of the time of year and if you plan on fishing it spring or fall, bring the long underwear.

Bridgeport is located on the U.S. Highway 395 corridor that parallels the eastern slope of the High Sierra from Bishop, California to Carson City, Nevada. Lodging and supplies are in Bridgeport with recreational vehicle

Dave Nolte surprised this 6 pound 'bow off Rainbow Point with Denny's Black Seal Bugger.

parks located along the lake. Primitive camping below the reservoir is available on the East Walker River.

For up to date fishing information, contact Ken's Sporting Goods in Bridgeport (760) 932-7707.

Bridgeport Reservoir Summary

SEASON: Opens last Saturday in April and closes the end of October.

TROUT: Rainbow and brown trout with some browns over 10 pounds. Average is 12 to 20 inches.

LAKE SIZE: Approximately 3,000 acres when full with the average depth about 8 feet.

RECOMMENDED FLY LINES: The intermediate when fishing the shallow upper end and a clear transparent line if the lake is flat. The area near the dam will fish best with a type II or III fast sinking line.

RECOMMENDED FLY PATTERNS: The Seal Bugger is good all season, A.P. Emerger and Stillwater nymphs in size 10 are good bets when the damsels emerge and the Callibaetis Nymph in size 10 or 12 is an excellent choice when fish are working the surface. Fish with aquatic insect patterns when the water warms. Minnow patterns, Seal Buggers and leech patterns will fish best early and late in the season.

BEST TIMES: Expect late May into early June prime time and again late September into mid October.

Butt Lake rainbows grow big as evidenced by this 27 inch male estimated at around 11 pounds.

BUTT LAKE
California

Butt Lake in northern California is one of the premier trophy fisheries in the state. Anyone who has hooked and landed one of these big trout will agree. Fed by nearby Lake Almanor, Butt is not your typical stillwater fishery. Minimal weed beds to explore, no shallow shorelines for trout to cruise and there are few aquatic insects to try and imitate. Make no mistake, there's plenty of food for these trout to eat and feeding is something they take seriously. For that very reason light tippets should be left at home if you hope to land one of these giants.

Butt Lake is both a rainbow and brown trout fishery with the average fish running 3 to 4 pounds, little guys compared to Butt's standards. Its reputation, and it is well deserved, is for big trout that live here. Rainbows from 8 to 12 pounds are fairly common with a monster weighing 17 pounds landed in the past years.

The secret to landing these big trout is timing. Prime time is when water releases from Lake Almanor activate the turbines. Pond smelt, a little minnow type forage fish, are trapped and funneled into the turbines and released into the narrow channel that feeds Butt Lake. Large numbers of these big trout have become accustomed to these water releases and will feed heavily on the dead and injured smelt that inherently drift down the channel. If you are here

This 25 inch female was taken in shallow water on Denny's Olive Chub Minnow and an intermediate line.

when these releases occur, the chances of hooking some of these overweight brutes is very good. So what happens between water releases? It is tough fishing, but not impossible. If other food sources are dominant enough to draw the trout's attention, you have a shot. Anglers using aquatic insect patterns will find it difficult to hook fish in Butt. Along with the pond smelt, leeches, crayfish and sculpins provide the large majority of what these big trout eat. Aside from the Hexagenia mayfly hatch, the trout concentrate little on aquatic insects.

Streamers imitating the pond smelt along with Leeches, Woolly Buggers and Seal Buggers offer the best odds of catching fish. If the Hex hatch is on, Jay Fair's Hexagenia nymph can be deadly when drifting it with the current.

Even though these fish are not leader shy, a soft tip rod is a good idea. Most fly lines will work, but you can fish your imitation up near the surface to look natural. A floating or intermediate line will get the job done. A 10 to 12 foot leader with a 3x tippet matches up well with these trophies.

Because there is no closed season, the best time for Butt Lake trout is March through June and again in the fall from late September through November. Mid-day fishing is usually slow, however, you can work it into the black as night fishing is legal.

Butt is a long and narrow body of water with standing dead timber, ideal cover for these big trout. Most of the lake is relatively shallow, 4 to 8 feet with both rocky and sandy bottom areas.

Because this lake is so dependent on water releases for active feeding, you will need to spend some time here to have a good shot at these tackle busters. It's a moody lake in between, but patience pays off.

Lodging and accommodations are available in Chester located on Lake Almanor. If Butt doesn't produce, Lake Almanor is a quality fishery in itself and produces big fish of its own.

To get there take Interstate 5 to Red Bluff and take the Highway 36 Exit. Drive 65 miles east to Highway 89. Turn left off of Highway 89 and go 6 miles to Butt Valley Road. The lake is 3 miles in on a gravel road.

Butt Lake Summary

SEASON: Open year-round.

TROUT: Rainbow and brown trout with fish up to 18 pounds possible. Average 3 to 4 pounds.

LAKE SIZE: Butt is a long narrow lake with an average depth of 4 to 8 feet.

RECOMMENDED FLY LINES: Floating and intermediate lines will work best. When fishing with minnow imitations, use the intermediate and work your flies just below the surface. You can explore the deeper regions with a type II full sink, but it is not as productive.

RECOMMENDED FLY PATTERNS: Minnow imitations to match the pond smelt will take the biggest fish. Jay Fair's Hexagenia mayfly nymph imitation works well when the Hex hatch is on. Insect patterns will take the small trout and a big fish on occasion.

BEST TIMES: With no closed season, I prefer March through June and late fall from the end of September through November. The key is to be there when the water releases are occurring.

CROWLEY LAKE
California

Crowley Lake, a high desert gem, is California's most popular and most productive fishery. (Dave Nolte photo)

On the eastern slope of California's High Sierras, a number of quality lakes blend nicely with the sagebrush in a high desert setting. Fed by numerous small streams along with the Owens River, Crowley Lake stands out as one of the state's best stillwater fisheries. Like most high desert lakes, Crowley is somewhat shallow, rich in nutrients, and is blessed with vast aquatic gardens capable of supporting prolific insect hatches and boundless food sources to grow large trout.

Kamloops rainbows averaging 16 to 20 inches form the bulk of the trout angler's catch, but trophy browns, some to 10 pounds, are not at all uncommon and monsters to 15 pounds occasionally are hooked as well as a few cutthroats. Sacramento perch have become prominent members of the lake's fishery and their fry comprise one of the staples in the diet of these large trout.

Anglers match up using a host of minnow imitations but it's the Matuka that fly fishermen use to imitate the perch fry. Body colors vary among the olives with medium olive tones the most popular. I've had my best luck using my olive Seal Bugger in size 8 fished in the top foot of the lake's surface.

There is an abundance of aquatic insects for trout to pick and choose from, most notably the Callibaetis and Trico mayflies along with caddis, chironomids, damselflies and dragonflies. Surface action can be intense at times, but usually the window of opportunity is short-lived. When it's happening, the Elk Hair Caddis, Pale Morning Dun and Adams along with the Para Duns and Thorax Duns in sizes 12 to 18 will match most of what appears on the surface. Small nymphs are effective as well with most of the standard patterns all well

Brown trout, although fewer in number compared to rainbows, have exceeded 20 pounds in years past.

This 5 pound rainbow, a survivor from the prior year's angling pressure, was taken on Denny's Olive Stillwater Nymph.

represented. When fish are showing, I've enjoyed consistent action by using my Stillwater, A.P Emerger, and Callibaetis nymphs in sizes 10 to 14. When fishing any nymph pattern, keep your fly in the top 2 or 3 feet. An intermediate or floating line will match most situations you encounter. The Matuka, my Seal Bugger, Woolly Bugger or Leeches can be effective fished up or down. More browns seem to be taken near the bottom and the rainbows closer to the surface. Uniform sink lines with varying sink rates are excellent for probing deeper water.

Megee Bay, the mouth of the Owens River, Alligator Point, Leighton Springs and Sandy Point are a few of the better areas for anglers to explore, although fish are often scattered throughout the lake during the spring months.

Crowley's trout are aggressive fighters that will often test your backing as well as your skill level. Tippets of 2x and 3x are common when fishing with the larger minnow type flies. When fishing smaller nymphs and leech patterns, 4x and 5x tippets work better.

The season begins the last Saturday in April and concludes July 31, although an extended season begins August 1 and runs until October 31. During the extended season, a two fish limit is in effect with an 18 inch minimum length and only single, barbless hooks with flies or lures allowed.

At 6,720 feet above sea level, Crowley chills down quickly regardless of the time of year. Wind, or a lack of it, is always a factor relative to presentation and mobility. There are no overnight accommodations at the lake, but there are plenty of places for campers to stay within a short drive from Crowley. Motel accommodations are available in Bishop about 30 miles south of the lake or in the Mammoth Lakes area about 20 miles north off U.S. Highway 395. For up to date fishing information, contact Fred Rowe, Sierra Bright DOT at (619) 934-5514.

Crowley Lake Summary

SEASON: Opens the last Saturday in April and closes the end of October.

TROUT: Rainbow and brown trout with an occasional cutthroat. The rainbows run up to 5 pounds and the browns up to 15. The average trout runs 2 to 3 pounds.

LAKE SIZE: Crowley has 45 miles of shoreline and is relatively shallow, averaging 8 to 10 feet.

RECOMMENDED FLY LINES: Although fly fishermen use them all, I prefer the intermediate line when fishing the top 6 feet and the clear transparent line for 7 to 12 foot depths.

RECOMMENDED FLY PATTERNS: The Seal Bugger has been the most consistent pattern for me. Olive or black work well all season. When trout are showing, I like my Stillwater, Callibaetis or A.P. Emerger nymphs in size 10 fished in the top foot of the lake's surface.

BEST TIMES: There isn't a bad time to fish this lake, but most fly fishermen concentrate their efforts in late September through October. The bigger trout, especially the browns, seem to be more available in early spring as well as the fall.

Davis Lake, rich and fertile, with shallow bays and fast growing trout, is one of California's best fly fishing lakes.

DAVIS LAKE
California

Located in the heart of the Feather River country, Davis Lake is really a reservoir situated about three miles above the little town of Portola. This is pine tree and sagebrush country where lakes are fertile and capable of growing large trout. In the fall of 1997 the lake was poisoned to eradicate pike and other non-native species illegally introduced. The lake was restocked in the spring of 1998 and with the fertile habitat still intact, trout to 6 pounds are available once again. Being a relatively shallow lake, Davis supports rich aquatic vegetation in its upper reaches.

It is home to both rainbows and browns although the browns are much fewer in number. Fish average 14 to 20 inches with 5 to 6 pound fish common. Years ago the lake supported a healthy population of cutthroat, but they have long since disappeared.

The best fly fishing is in the upper regions of the lake along with the shallow bays and coves where weed beds provide both feed and cover. May through July are prime months to fish and again in the fall when the trout move back into shallow water to feed.

Davis Lake rainbows are treated to a wide variety of trout foods accounting for large numbers of big trout.

Davis supports intense hatches of damsels, midges and Callibaetis mayflies to go along with leeches, scuds and dragonfly nymphs. By mid-June damsels become a major food source for trout providing anglers with an excellent opportunity for consistent action. Trout will move into the weedy shallow bays during the early morning hours to exploit the little damsel nymphs. Try not to miss it if you can help it. A floating or intermediate line, along with Dave Takahashi's Little Olive Damsel Nymph is a deadly combination. Of course, it takes a good presentation to make it happen. The other staples in the diet of these stillwater trout are the Callibaetis mayflies and midges, which come off throughout the season. Anglers can use Hare's Ears and my Callibaetis nymphs in sizes 12 to 16 to match the mayflies and a Midge Pupa in sizes 16 to 20 when fish are taking midges.

In the spring of 1997, I was tubing one of the upper bays and was astonished at the number of adult orange-bodied midges covering the surface. Surprisingly, not a single trout stuck its nose out to eat one. These were not tiny bugs either. A size 10 or 12 matched the insect perfect, but without any surface activity I continued to probe the weedy areas with my Stillwater Nymph and Seal Bugger. I struggled until I saw a big trout finally rise and eat one of the countless midges laying on the lake's surface. So what was wrong? The problem was simple, I was fishing barren water. The trout hadn't moved in to feed yet. As more fished moved in, the little insects became easy prey for the hungry trout. In the two hours that followed, I enjoyed outstanding action for rainbows 18 to 23 inches on my Cinnamon Callibaetis Nymph. That fly proved to be the perfect match for the reddish-orange-bodied adult until the wind ended the surface activity a couple of hours later. It was sight fishing at its best.

Dry fly fishing can be productive as well as exciting during the summer months when trout decide to feed on top. You will have to match up, but the action is furious while it lasts.

Davis is a lake made for car top boats and float tubes. Wind is seldom a limiting problem, but these trout can be spooky without some form of cover.

Floating and intermediate lines will fish the shallow bays best, but don't be without a clear or transparent line when fishing windless periods. Without

Fly fishermen score big at Davis in late June when the rainbows, like this 3 pound male, follow migrating damsels into shallow water to feed.

them, you will end up spooking far more fish than you will ever hook. Launch ramps and camping are available around the lake with lodging and supplies available in Portola. For up to date fishing information, contact Dave Takahashi at Grizzly Country Store (916) 832-0270 located on the lake.

Davis Lake Summary

SEASON: Opens the last Saturday in April and closes the end of October.

TROUT: Rainbow and a few browns with the average trout 16 to 20 inches. Fish to 6 pounds are not uncommon with an occasional larger brown.

LAKE SIZE: When full, the lake is about 4,000 acres with numerous bays producing the best bite.

RECOMMENDED FLY LINES: The intermediate is the best line for this lake, but the floating and clear transparent lines have their time and place.

RECOMMENDED FLY PATTERNS: Seal Buggers are good early and late in the day or season. When insects are hatching, I prefer my Callibaetis Nymph or A.P. Emerger in sizes 10 and 12 and my Stillwater Nymph when damsels are hatching.

BEST TIMES: This lake produces well all year, but you must adjust the pattern to those insects that are available at the time you are fishing. Summer and fall are fairly consistent for fly fishermen.

NOTE: In July, 1999, pike have once again been discovered in Davis Lake leaving disposition of the lake's fishery in doubt.

EAGLE LAKE
California

Eagle Lake, home of the Eagle Lake Rainbow, is a favorite of stillwater anglers from October through December.

Eagle Lake is home to the famous Eagle Lake rainbow, a strain of trout once believed to be extinct. Fortunately, they're still thriving here. Catching these big trout isn't difficult, locating them is. With countless miles of shoreline to cover, anglers unfamiliar with Eagle and her moods tend to suffer the most. Although some depths are over 90 feet deep, it's the shallow shoreline areas that produce for fly fishermen.

These rainbows are hardy fish thriving on a diet of scuds, damselflies, dragonflies, leeches and small forage fish. What you won't find are extensive hatches of mayflies, caddis or midges for anglers to imitate.

Eagle Lake opens to angling the Saturday preceding Memorial Day and closes December 31. It's the fall months, October through December, when fly fishing reaches its peak. Late November through December can be extremely cold with freezing temperatures the norm. But, this is a special fishery that only the dedicated angler will pursue and it's worth it. During this period, water temperatures become cold enough to move scuds by the millions into the shallow water and the trout are right behind them frequenting these areas early and late in the day.

Eagle's rainbows average 2 to 4 pounds in the spring and add a pound or two by the fall. Fish to 10 pounds are not as plentiful as in past years, but they are there. Five to 8 pound fish are much more common.

Productive patterns include Jay Fair's Crystal Buggers with fiery rust hackles, brownish-orange Leeches, my Seal Bugger and Stillwater Nymph which I'm partial to because of its resemblance to the scuds. Leaders of 10 to 12 feet tapered to a 4x tippet are good choices as most of these trout are not leader shy.

Three fly lines are necessary to fish Eagle effectively: an intermediate, the clear transparent type II and a floating weight forward line. When trout cruise the shallows, most takes will occur in 1 to 3 feet of water.

Because of Eagle's size, a boat with at least a 15 horsepower motor is a must to reach the more productive areas. Trout cruise in and out along the rocky shorelines and anglers who beach their boats and wade will be the most productive. Tules line parts of the western shoreline and anglers fishing from float tubes or boats casting tight against the tules will find some healthy fish waiting for them. This, however, is the only manner in which I find fishing from a floating device at this lake productive. My close friend Jay Fair taught me many lessons on where and how to fish this lake as well as other lakes from shore. His knowledge and methods for fishing this huge body of water are second to none. If you are first timer, look him up.

Winds are a common occurrence on the lake but trout use the riffles as cover when feeding in shallow water. Caution when boating is always a wise option especially for inexperienced anglers.

U.S. Forest Service campgrounds are located at the south end of the lake along with a launch ramp. Spaulding on the western shore also provides a launch ramp with more facilities as well. Motel accommodations are available in Susanville. For up to date fishing information contact Jay Fair Guide Service (530) 825-3401.

Eagle Lake rainbows will average 3-5 pounds by fall with fish to 10 pounds possible.

The late Hap Scollan with a 4 pound male taken on Denny's Stillwater Nymph.

Eagle Lake Summary

SEASON: Opens the Saturday preceding Memorial Day and closes December 31.

TROUT: Eagle Lake rainbows average 2 to 4 pounds with a fair number of 5 to 8 pounders fairly common.

LAKE SIZE: Eagle Lake is huge and varies between 29,000 acres when full and as low as 16,000 acres during drought years. Average depths vary depending upon where you fish, but are unimportant to fly fishermen since the best angling with flies is the shoreline areas or tight to the tules in late fall.

RECOMMENDED FLY LINES: The intermediate is best, but a floating line works well when fishing shallow water.

RECOMMENDED FLY PATTERNS: I've landed these rainbows on numerous patterns, but none better than my Stillwater Nymph in size 10.

BEST TIMES: The best fly fishing begins in late October and continues until the lake freezes in December.

Heenan Lake, a broodstock cutthroat fishery, supports huge numbers of 2-5 pound cutthroats with fish to 10 pounds possible.

HEENAN LAKE
California

In 1981, I had an opportunity to fish California's Heenan Lake as part of a study to determine if it was feasible to open the lake to the public. Since the lake is utilized as a Lahontan cutthroat brood stock fishery supplying eggs to hatcheries around the country, the lake had never been open to angling before. Anglers waited anxiously for the decision and when it came, the cheers went up.

It was no surprise that special regulations would be imposed to protect the sensitive fishery. Based on recommendations from the Department of Fish and Game, the lake was opened in 1984 for fishing on Friday, Saturday and Sunday beginning with Labor Day weekend through the last Sunday in October. The fall dates were adopted to reduce the high probability of fish mortality during the hot summer months.

The average cutthroat in Heenan runs 16 to 20 inches, but fish from 6 to 8 pounds are common and cutts to 10 pounds are landed occasionally. But much bigger fish are there although fewer in number as evidenced when these fish are artificially spawned in the spring every year.

Heenan is a small lake with 120 acres of water, is rich in nutrients with algae forming during the summer months. One small tributary feeds this relatively shallow lake with the deepest water about 18 to 20 feet.

An abundance of aquatic insects provide exciting sub-surface action for stillwater fly fishermen.

Insect hatches are pretty limited by fall with a few Callibaetis mayflies, midges and an occasional damselfly showing up. These fish are not too picky about food preferences. Most standard nymphs in sizes 10 to 14 along with Leeches and Woolly Buggers in sizes 8 or 10 are excellent choices. If the water is off color a bit, try the darker colored flies. I usually do quite well with a black Seal Bugger regardless of water quality.

A number of lines will work at Heenan and of course, with a shallow water fishery, I prefer the intermediate to start. If the cutthroats aren't in the top 4 feet, I'll switch to a uniform sink II and probe the next 6 feet. Use a floating line if surface activity is evident. Since these trout are not leader shy, 9 to 12 foot leaders with tippets of 4x or 5x to match pattern size is always a good choice.

Most areas of the lake hold some fish, but as seasonal changes occur so will their holding areas. I prefer to fish just off the weed beds near the inlet and the upper end of the lake just off shore, but always near cover.

Daytime temperatures are usually very comfortable but nights can be frosty in the fall so dress accordingly. Current regulations require all trout to be released unharmed using either flies or lures with single barbless hooks. There is a $3.00 daily fee to fish the lake and permits can be obtained in near-by Markleeville.

Heenan can be reached off Highway 89, referred to as Monitor Pass Road about 8 miles south of the sleepy little town of Markleeville. There is no camping allowed at Heenan, but there are unimproved spots along the East Carson River a few miles away. Accommodations and supplies can be purchased in Markleeville. Because this is a sensitive fishery, check the regulations annually before heading to Heenan.

Heenan Lake Summary

SEASON: Open Fridays, Saturdays and Sundays beginning Labor Day weekend through the last Sunday in October.

TROUT: Lahontan cutthroat with fish to 8 pounds. Average fish runs 16 to 20 inches.

LAKE SIZE: Heenan covers about 120 acres when full with an average depth of 8 to 10 feet.

RECOMMENDED FLY LINES: The intermediate is perfect for this lake when fish are feeding.

RECOMMENDED FLY PATTERNS: The Black Seal Bugger when searching and the Stillwater Nymph and A.P. Emerger in size 10 if fish are working near the surface.

BEST TIMES: Any day it is open concentrating on early and late periods of the day.

A Kirman Lake male brookie in full spawning dress.

KIRMAN LAKE
California

Located 20 miles north of Bridgeport, Kirman Lake is perhaps the best trophy brook trout fishery in California, and certainly among the better ones in the West. This is a hike-in lake about 3 miles off Highway 108 just west of the junction with U.S. 395. It's mostly an uphill climb, but it is a gentle slope and not difficult to negotiate.

Before Kirman was really discovered by fishermen, stories of 4 to 6 pound brookies were well documented among local anglers. Understandably, rumors of trophy brook trout were apparently too difficult to keep a secret. In time, the trail to Kirman became as well defined as heron tracks along a river's edge. Those who have made the trek and landed one of these beautiful trout know what a delicate lake it is. Part of the promise here is more than just big brook trout — the lake also harbors big cutthroat. However, the cutts aren't the primary focus of those who fish here. It's those big, beautiful brookies that bring fishermen up the trail.

Kirman, pronounced Carmen, is fairly small, about 60 acres which allows anglers to spend less time locating fish. Like brown trout, brookies are structure huggers and are usually found in deeper water until time to feed. Most big trout, brookies included, can be temperamental, especially when spring and fall weather patterns become subject to change. This was the case when I last fished Kirman, just as the ice was coming off. Although the hike was adventurous at times, trotting through deep snow in places, it was well worth the effort. The day started a bit slow, but improved dramatically about 9 in the morning. Fishing eventually slowed again in the early afternoon as a weather front approached. I didn't land any cutthroat, although I saw several around 4 or 5 pounds, but most of the brookies I did fool went 15 to 20 inches ranging from 2 to 3$^1/_2$ pounds.

The trout in Kirman have always been partial to olive and spruce Matukas along with dark Leech and Woolly Bugger type patterns. Personally, I did well with my black Seal Bugger in size 8. When that fly slowed, I used my olive Callibaetis Nymph in size 10. That worked well for trout working near the surface. I'm quite sure most standard nymphs will produce here, but keep them on the olive and black side.

The primary food source found in this lake are scuds and since most of these big trout prefer to feed below the lake's surface, anglers need to explore the lake with sinking lines. I used my intermediate when the trout were shallow, but a number II full sinking line to work near or off the bottom would be a better choice during the late spring and summer months.

Leaders of 9 to 12 feet are a good choice as neither trout species is leader shy. I found tippets of 4x good for spring and again for fall fishing, but plan on sizing down to 5x when the conditions require it.

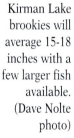

Kirman Lake brookies will average 15-18 inches with a few larger fish available. (Dave Nolte photo)

The best time for Kirman is spring and again in the fall. Trout are shallow enjoying higher oxygen levels after ice out, but will move into deeper water as the water warms. Summer time is usually on the slower side, but fish still have to feed and will return to the shallower water along the shoreline areas early and late in the day. Fall fishing is a special time at Kirman. The brookies are in full spawning dress and fishing pressure is generally on the light side. This is an excellent time to probe the shallow water where most of the bigger trout can be found feeding a few feet below the surface.

Most of the fishing on Kirman is done from a float tube or pontoon type boat. Wading is not as effective during the summer months.

The fun begins on the last Saturday in April and the lake remains open through October. There are no improved campgrounds at Kirman, but you will find many situated along Highway 395 where the Little Walker and West Walker rivers run parallel to the highway. Lodging and supplies can be found in Bridgeport.

Remember, these higher elevation lakes can be subject to sudden changes in temperature when weather fronts move through. Fly fishing can be exceptional or an adventure depending on what Mother Nature has in store. One thing's for sure: if you land one of these big trout, you will want to capture the moment on film so don't forget the camera.

For information about fishing conditions call Ken's Sporting Goods in Bridgeport (760) 932-7707.

Kirman Lake Summary

SEASON: Opens the last Saturday in April and closes the end of October.

TROUT: Brook and cutthroat with the average brook trout 2 to 3 pounds and up to 6. The cutthroat run 3 to 4 pounds with fish to 8 pounds possible.

LAKE SIZE: About 60 acres with an average depth of 8 to 12 feet.

RECOMMENDED FLY LINES: An intermediate is best for spring and fall fishing in shallow water and a fast sinking type II or III better for summer or anytime you want to probe the depths.

RECOMMENDED FLY PATTERNS: Black Leeches and Seal Buggers in size 8 are good all year. An old standby has been an olive Matuka. When trout are surface feeding, I prefer my olive A.P. Emerger or Callibaetis Nymph. All standard nymph patterns should work in black and olive colors in sizes 10 or 12.

BEST TIMES: Spring and fall are best and are fairly consistent.

This fall spawner was fooled by Denny's Tan Callibaetis Nymph fished in the top few inches on an intermediate line.

MARTIS CREEK RESERVOIR
California

There is a growing population of stillwater anglers who believe this little reservoir may be the best fly fishing lake in California. Like most stillwater fisheries, there are days when no one would argue the point. This may be relevant to who you talk to on a given day, but it certainly must be considered as one of the best. Martis has some big trout that make the challenge well worth the effort. Like most lakes, it can be temperamental if you aren't familiar with her moods. But, if you know when and how to approach these big fish your experience can be extremely productive.

Martis was built to help control potential flooding in the area and has since developed into a top trophy fishery. The lake covers about 75 acres when full, and is rich in aquatic insects that provide the bulk of what the rainbows, browns and Lahontan cutthroat eat. The 'bows average 14 to 18 inches with 4 to 6 pound fish possible, while the cutts, although fewer in number, run 16 to 20 inches with fish to 6 pounds on occasion. The browns are probably the biggest challenge with fish over 10 pounds lurking around cover and around the many springs that exist within the lake.

A multitude of hatches keeps anglers guessing and the trout feeding. Midges, Callibaetis mayflies, damselflies and scuds in sizes 10 to 14 are the primary food sources for the anglers to imitate. Spring time means fishing attractor patterns such as Seal Buggers, Woolly Buggers, Leeches and streamer flies like the Matukas and Zonkers in sizes 6 to 10. When summer temper-

The author with a 6 pound rainbow taken on his stillwater nymph in shallow water. (Dave Nolte photo)

atures arrive, dry fly fishing increases. Anglers can catch fish using Parachute Adams, Griffith's Gnat, Blood Midges or any pattern that matches the hatch at the time. When damsels are active, fish olive damsel nymphs in size 10 or 12. I do quite well during the damsel hatch with my Stillwater Nymph fished on an intermediate line and a hand twist retrieve. If any surface activity occurs, I'll switch to my Callibaetis Nymph and fish it in the top 6 inches with a slow hand twist retrieve or a slow pull and pause retrieve. This is sight fishing at its best and very consistent.

You will need longer leaders from 12 to 15 feet and 5x or lighter tippets to play the game, a game anyone can play. Other patterns that have proved to be effective include Zug Bugs, Hares Ear and Pheasant Tail Nymphs in sizes 12 to 16. The Chironomid Pupa and a Blood Midge fished as an emerger are also good. Long leaders of 15 to 20 feet are necessary when fishing emergers or dries as these big fish will spook quickly on shorter leaders. Tippets may have to get down to 6x to be effective.

The most productive areas are the upper end of the lake and around the channels entering the lake or near the springs during the summer months. The lake fishes well all season beginning in May although August into mid-September can be slow except for early and late in the day. Fall fishing isn't as productive in terms of numbers, but it can be for size.

Martis is an artificials only, catch and release fishery with single barbless hooks allowed.

Don't expect much solitude here as the lake's reputation precedes it. That equates to a lot of fishing pressure but the lake seems to hold up well in spite of it. This is an ideal lake for float tubes or small boats and prams, although shore fishing can be productive at times.

The lake is located about 3 miles southeast of the town of Truckee on Highway 267 that leads to Lake Tahoe. There is a campground at the lake and lodging and supplies are available in Truckee. For fishing information or guide service contact Randy Johnson's Guide Service (530) 525-6575, or Mountain Hardware (530) 587-4844.

Martis Creek Reservoir Summary

SEASON: Opens the last Saturday in April and closes November 15.

TROUT: Rainbow, brown and Lahontan cutthroat. Rainbows average 14 to 18 inches with fish to 6 pounds possible. The browns will average 16 to 20 inches with fish to 10 pounds while the cutthroats consistently average 15 to 20 inches with 6 to 7 pounders available.

LAKE SIZE: Martis is about 75 acres when full with lots of 6 to 8 foot shoal areas to fish.

RECOMMENDED FLY LINES: Spring time and fall fishing can be done with floating or intermediate lines. When trout go deep during the summer months, faster sinking lines such as a Uniform Sink #2 or Cortland's clear transparent line will work just fine.

RECOMMENDED FLY PATTERNS: You can empty the box on this one as lots of patterns will work. Concentrate on the depth the fish are feeding. I prefer my Seal Bugger for early spring and late fall and my Callibaetis Nymph or A.P. Emerger in olive when fish are showing. When damsels are hatching, I use my Stillwater Nymph in size 10.

BEST TIMES: Spring and fall are almost always good. During the summer months, these fish will go deep, but will return to the shallows or just under the surface to feed.

McCOY FLAT RESERVOIR
California

Located about 15 miles north of Susanville and some 30 miles south of Lassen National Park, McCoy Flat Reservoir supports huge numbers of trophy rainbows. Still, this lake receives very light pressure from stillwater anglers. When a lack of water isn't a problem (the lake has been drained seven times in the past twenty five years), McCoy can be a dominant fishery equal to any trophy lake found in the state. When the lake level is maintained due to sufficient snow pack, these trout can reach double digit proportions within 3 to 4 years. No wonder anglers get excited.

Like all shallow, nutrient-rich lakes, lack of oxygen is a problem in the winter and fish loss can be detrimental. During cold winters, McCoy experiences this natural occurrence. Fortunately, the reservoir is fertile, somewhat rich in aquatic weeds and very shallow with an average depth of about 7 feet when full. That allows the trout to grow big. Add in intense hatches of Callibaetis mayflies, damselflies, midges and dragonflies and it's no wonder why this lake sustains itself even in down years. But the real secret to fast growth of these trout are the scuds and ghost shrimp that seem to reappear even during the years the reservoir is drawn dry.

These rainbows are as tough a fighting trout as I have hooked in any lake in the West. Maybe that is a testament to their diet, or perhaps the strain of trout, perhaps both. If you hook one, you can bet they will test your tackle and your skills.

McCoy Flat Reservoir is a shallow lake rich in aquatic insects that grow big trout quickly.

Although McCoy is primarily a rainbow fishery, brook trout help spice the action. These trout grow so fast it is hard to establish an average size. The rainbows are stocked when they're 8 to 10 inches long and grow to 18 to 20 inches in just a year. Fish from 8 to 10 pounds are not only possible, but are fairly abundant while brookies will reach 5 to 6 pounds after about three years in the lake.

These trout are aggressive feeders and move around the lake holding in the old river channels and near the dam when water becomes a problem. The lake has little to offer the trout in way of cover, so reading the water for the best areas to fish can be a major challenge for anglers. Fly fishermen unfamiliar with McCoy tend to struggle more than those who spend quality time becoming familiar with her trout and their moods. Having the skills to catch trout is one thing, but knowing where and when to catch fish in this lake is even more critical.

McCoy fishes best when insect hatches begin appearing by late May or early June and can offer consistent action well into the fall months. Small nymphs in size 10 to 16 are best with the Prince Nymph, Pheasant Tail and Hares Ear nymphs local favorites. I found my best action using my tan Callibaetis Nymph in size 12 when surface action was obvious and my Stillwater Nymph in size 10 when damsels started appearing. There are forage fish in McCoy, but Leeches, Woolly Buggers and my Seal Bugger have done very well in olive with olive or burnt-orange tails or black with black and burgundy tails. That is especially true early and late in the season when food sources are not as abundant.

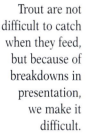

Trout are not difficult to catch when they feed, but because of breakdowns in presentation, we make it difficult.

Because McCoy is so shallow, there is no need for fast sinking lines. I use only my intermediate or the clear transparent line especially when conditions dictate. Along with the clear transparent fly lines, long leaders of 12 to 15 feet are a definite advantage in flat, clear water. Some of the best fishing will occur when these big trout cruise shallow water and you will need a long leader to keep from spooking them. As for tippets, it's important to match the tippet with the fly, but as a rule I usually fish 4x fluorocarbon tippets especially in clear, shallow water.

Early morning and late evening are prime times during the summer although fall is typically the best time of the year to fish McCoy. I have not done well personally in the early spring when temperatures are on the cold side and the reservoir is full. This tends to scatter fish and it's time consuming to look for feeding trout.

Wading is good when these fish are cruising, but a boat or float tube will allow you to reach more fish. You will need these options when searching or exploring and especially when the water warms and trout hold in the old streambed channels.

There are no campgrounds at McCoy, but primitive camping is allowed. For lodging and supplies Susanville, located about 12 miles south, will cover your needs.

McCoy can be temperamental so don't get discouraged. Spend some time and learn this lake. It is worth it.

Barney Scollan's four year old rainbow measured 24 inches with an estimated weight of 7 pounds proving McCoy is an intensely rich lake.

McCoy Flat Reservoir Summary

SEASON: Opens the last Saturday in April and closes the end of October.

TROUT: Rainbow and brook with the average rainbow running 3 to 6 pounds with fish over 10 pounds. Brookies will average 14 to 16 inches with fish to 5 pounds possible.

LAKE SIZE: McCoy has about 95 surface acres when full with an average depth of 7 feet.

RECOMMENDED FLY LINES: The intermediate or floating is all you need.

RECOMMENDED FLY PATTERNS: Fish Seal Buggers early and late in the season. I prefer my Stillwater Nymph when damsels are showing and I like my tan Callibaetis Nymph or A.P. Emerger when trout are feeding near the surface.

BEST TIMES: Early or late in the day during summer months can be very good but more big trout are taken in the fall when fishing pressure is down and big trout cruise shallow water.

Colorado, with an abundance of stillwater fisheries, offers a variety of trout species and scenic beauty. (Dave Nolte photo)

Chapter 3

Colorado

When anglers refer to fly fishing in Colorado, places like the Frying Pan, South Platte, Roaring Fork or the Gunnison rivers understandably come to mind. What you don't hear about are such places as Spinney, Delaney Butte, Elevenmile or Blue Mesa. All of these are lakes or reservoirs and there are many others that harbor big trout, some of which grow to double-digit size. Not all of the lakes in Colorado focus on the trout; pike and tiger muskies are getting lots of attention from resident anglers.

In some of Colorado's trophy waters they have mixed pike with the trout. Whether planned or by accident, the trout have suffered in every instance. Take Spinney Mountain Reservoir for example. In my opinion it is the best stillwater fly fishing lake in the state, but the pike are on the increase again. In time, the problem of over population of these extremely aggressive fish will have to be dealt with if Spinney Mountain is to maintain its legendary trout status.

When you think of Colorado you can't help but think of the Rocky Mountains, which are not only a scenic attraction, but are sprinkled with mountain gems where rainbow, cutthroat and brook trout thrive.

During my college days I spent a summer working, playing ball and fishing the Grand Mesa Lakes near Delta. When I wasn't hitting a baseball, I usually had a line in the water exploring these pristine lakes. The Mesa still holds some quality stillwater fishing, but like most of our better fisheries around the country, it's not quite the same today. Fishing pressure is on the increase in this state and you really have to find a hidden gem to get away from the crowd. Fortunately, you will feel and experience more pressure during the summer months than spring or fall when stillwater fishing can be in its prime.

This is a beautiful state and well worth the time spent exploring its natural scenic wonders. My advice is to spend some time, visit, admire and enjoy. And bring a fly rod.

NORTH DELANEY BUTTE LAKE
Colorado

Terry Bridgeman with an 8 pound brown from North Delaney Butte Lake. (Terry Bridgeman photo)

Many veteran anglers consider this lake in the same class as Spinney Mountain Reservoir, one of Colorado's best trophy fly fishing lakes. I can't disagree with that assessment. North Delaney Butte is by any definition a truly outstanding fishery supporting brown trout that average 14 to 20 inches, but with a good number of big fish over 10 pounds. Rainbows and cutthroats contribute to the action and although fewer in number, fish from 5 to 6 pounds are not uncommon.

North Delaney is a high desert lake in every sense of the word. Situated in sagebrush country above 7,000 feet, the lake is about 150 acres with heavy concentrations of weeds and aquatic insects. Anglers do well imitating damselflies and caddisflies from mid-June through August. Midges, scuds and leeches are available throughout the season. The most popular patterns include olive damsel nymphs in sizes 10 or 12, Midge Pupa in sizes 14 to 18, caddis dries such as Traveling Sedges and the Goddard Caddis in sizes 12 to 16, along with Leeches, Seal Buggers or Woolly Buggers in sizes 8 or 10. Black, brown and olive are primary colors, but sunlight always is a factor regarding color definition. Because of the many weeds, anglers need to work the openings between the masses or look for fish cruising the edges.

A floating or intermediate line will do the best job when imitating the natural insects available to these trout, especially if you are fishing in and around the weed beds. Sink tip lines do a good job when imitating ascending mayflies, especially using a slow retrieve.

Weather is often a factor for anglers here. It also has a tendency to put fish off their normal bite. Expect sudden windy periods and stormy situations spring and fall.

The lake does not lend itself to wading. You are better off fishing it from a boat or float tube.

South and East Delaney lakes are sister lakes located only a short distance away and are about the same size, but a notch below North Delaney in terms of quality. However, any of the three lakes can be the better fishery on any given day.

Camping is permitted on all three lakes, but is unimproved. Lodging, supplies and fishing information are available in Walden, a short distance away.

To get to Delaney Butte lakes, take the access road just south of the little town of Walden. Go 10 miles and turn right onto a good gravel road. Follow the signs to all three lakes.

This 3 pound brown was taken in late May on a black Seal Bugger from North Delaney.

North Delaney Butte Lake Summary

SEASON: Managed as Gold Medal Water, North Delaney Butte Lake is open year-round. Regulations call for flies and lures only with a 2 fish bag limit. Browns between 14 and 20 inches must be released.

TROUT: Smaller browns, rainbows and cutthroats share the space with the browns reaching double-digit sizes. The rainbows and cut-throats will average 14 to 20 inches but fish to 5 and 6 pounds are possible.

LAKE SIZE: Delaney has about 150 acres with lots of shallow weedy areas for fly fishermen to explore.

RECOMMENDED FLY LINES: Intermediate or floating lines are best for fishing the shallow areas or when trout are feeding sub-surface. I prefer a clear transparent line for flat conditions or when I want to fish 6 to 12 feet under the surface.

RECOMMENDED FLY PATTERNS: Seal Buggers and Leeches are good searching patterns during the spring and fall or early or late in the day. They also fish well when the browns are near the bottom in 10 to 12 feet of water. But, the most prolific pattern for me is my Stillwater Nymph, which matches the damsel hatches during the summer. My olive Callibaetis Nymph is a perfect match during insect hatches especially when adult midges are on the surface.

BEST TIMES: The bigger browns are taken early spring and late fall. Summer action is relative to conditions, but if fish are working, you can catch them anytime with the right line, pattern and retrieve.

Spinney
Mountain
Reservoir is
capable of
growing big
trout like this
5 pound
rainbow.

SPINNEY MOUNTAIN RESERVOIR
Colorado

Nestled in an open mountain meadow at an elevation of 8,600 feet, Spinney Mountain Reservoir interrupts the South Platte River in Central Colorado and spreads across approximately 2,500 surface acres. Considered by many veteran anglers as the state's best trophy fly fishing lake, Spinney's reputation is based on size and, when conditions are right, numbers as well.

Browns from 5 to 10 pounds and cutthroats from 4 to 8 pounds used to dominate the fishery, but are now the minority species. Rainbows, some of which run up to 10 pounds, are the dominant trout. The average fish runs 3 to 5 pounds. Northern pike, once eradicated from the lake, are on the comeback and that's not good from the trout fisherman's perspective. It's quite a scene when the planters are stocked and the pike are just salivating with jaws wide open. These big trout sustain themselves on a fish and leech diet with aquatic insects serving as snack food in between meals. Starting in late spring and lasting through early October, hatches can be intense with Callibaetis mayflies and damselflies providing the bulk of the action.

During a late September day a few years back, I was fishing Spinney with my close friend and associate Dave Freel. We were greeted with the usual high winds and heavy chop. We were both fishing intermediate lines and Seal Buggers and enjoying moderate success. Nothing huge, but six or seven rainbows from 3 to 5 pounds. It was almost noon when the wind died and the

Callibaetis began emerging. It was like waving a red flag. I was fishing a point off the bay near the entrance when trout began showing everywhere. Switching to my Callibaetis Nymph, it was some of the best sight fishing I've enjoyed. After a couple of hours, I landed 14 fish from 3 to 7 pounds, all thick bodied, healthy rainbows. Interestingly, the fish were feeding within 15 feet of shore in about 2 to 4 feet of water and the only anglers within sight were fishing out in the middle of the lake. Let that serve as a reminder that the greatest food-bearing region in any lake is in shallow water.

Crowds are seldom a problem here and although Spinney can be moody at times, expectations run high even when anglers have to work for their fish. One thing about this reservoir: when you get one on, chances are good it's going to be big. Because of their food preferences, large minnow imitations, Seal Buggers, Woolly Buggers, Leeches and Woolly Worms in sizes 4 to 8 are all effective on these trout. Small nymphs, like my Callibaetis Nymph, A.P. Emerger and Stillwater Nymph, have proven deadly when matched with intermediate lines and a slow, hand-twist retrieve. I'm sure a wide variety of popular nymph patterns would be just as effective when worked with the same system.

Depending on the conditions present, light tippets are a bad choice. Use 3x to 4x fluorocarbon tippets and you'll be fine. Long casts are a necessity if you want the best chance for hooking these big fish. Fast, full sinking lines with a variety of sink rates will allow you to explore the depths, but as I have said, I've found the majority of fish hooked were on intermediate and the clear lines in less than 6 feet of water.

Unlike small trout that need to feed constantly, these big fish may feed once or twice a day, or sometimes every other day. When they're on the prowl they usually will show at the mouth inlets, weedy areas and shallow bays to hunt for their food.

Early morning is prime time for Spinney's big rainbows.

Big browns are possible on Spinney, but you have to get up before they do.

Spinney Mountain Reservoir Summary

SEASON: Managed as a Gold Medal Water, Spinney Mountain is open usually from about April 15 to early December. Once it freezes, it closes to fishing. Flies and lures only are the rule with a trout limit of one fish over 20 inches. Pike have a 10 fish bag limit, only one over 34 inches. Spinney is a Colorado State Park with a $4 a day entry fee required.

TROUT: Rainbow, brown and some cutthroat with the rainbows and browns running up to 10 pounds and the cutthroats ranging from 4 to 8 pounds.

LAKE SIZE: Spinney is roughly 2,500 surface acres of water with lots of weedy shallow shoreline areas.

RECOMMENDED FLY LINES: I use only the clear transparent or intermediate lines here regardless of time of year.

RECOMMENDED FLY PATTERNS: I prefer my Seal Bugger in olive and burnt-orange hackle for probing the weedy areas in the spring and fall. My Stillwater Nymph worked great when damsels were hatching and I used my Tan Callibaetis Nymph when I saw fish working near the surface.

BEST TIME: Early spring is hard to beat with fall very good for big browns. During the summer and early fall, insect hatches bring these big trout up near the surface providing some exciting action.

Idaho—Big trout like this brown is what Idaho fly fishing is all about.

Chapter 4

Idaho

Idaho is a state known for some of the best fly fishing waters in the United States. Some of its rivers are legendary, but Henry's Lake in the northwest corner of the state ranks as one of the best stillwater fisheries anywhere and Lake Pend Oreille still holds the record for the largest rainbow ever landed on a rod and reel. If there is a negative, it is a lack of trophy fly fishing lakes although it is blessed with an abundance of good rivers, streams and big deep lakes where trollers score big.

The most exciting lake in the state just might be Pend Oreille. Rainbows over 20 pounds are consistently landed each year and on occasion, monsters exceeding 30 pounds are hooked, but seldom landed. These fish are taken on flies but are trolled for, not cast to with fly rods. Late fall and early winter (October through November) are the best times for challenging these big trout. Two other quality lakes in the area are Priest and Coeur d'Alene. Backing up the Snake River, Palisades Reservoir is an exceptional brown trout fishery but you need to learn her secrets as she is often moody. Daniels, Mackey and Magic reservoirs can be very productive, especially during the fall months and are well worth exploring.

Idaho offers some excellent steelhead action as well and for something different, there are very good kokanee and black bass fisheries waiting to be explored. This is a beautiful state and if you take the time to visit its many scenic wonders and recreational opportunities, you will not be disappointed.

HENRY'S LAKE
Idaho

Idaho's Henry's Lake remains one of the best stillwater fly fishing lakes in the West.

Situated in a picturesque valley only 20 miles from Yellowstone National Park, Henry's Lake has to be considered one of the premier trout lakes in the United States. Shallow, weedy and insect rich, Henry's not only supports big trout, but lots of them. Cutthroat and brook trout are taken up to 6 pounds and cutthroat-rainbow hybrids average 2 to 6 pounds with several up to 17 pounds landed in past years. These hybrids are aggressive, powerful fish that will test your skill and your backing — if you can hold them. It's not unusual for anglers who are familiar with the lake to land upwards of 50 to 60 fish a day over 17 inches. Henry's does not fish well for the dry fly addicts, but is exceptional for anglers using nymphs.

Most food forms that we find in stillwaters are not only available, but are present throughout the season even though these trout can be selective at times. The key is to find what they are eating and stay with it until they either switch or go off the bite which is rare.

Damselflies are one of the major insects for anglers to imitate. They will begin showing in July with midges and mayflies making appearances on and off all season. Scuds, dragonfly nymphs and forage fish balance out the trout's diet. Patterns should include olive damsel nymphs and scuds in sizes 10 to 14, Seal Buggers, Woolly Buggers, Leeches and minnow imitations in sizes 6 or 8 with olive and black as the predominant colors. I have enjoyed outstanding success using my Stillwater Nymph and an olive body/fiery rust saddle hackle Seal Bugger in June as well as September and October.

Because of the numerous weed beds and shallow water, an intermediate line does an excellent job of presenting your fly. But there will be many situations when the transparent and faster sinking lines are necessary. Some of the biggest trout in Henry's hide in the deeper pockets around weedy columns and you need to reach them to catch them.

Depending on the stiffness of your rod and the size of the pattern you're fishing, use either 3x or 4x tippets on a 10 to 12 foot leader. Most retrieves will catch fish, but I feel the slower pulls are more consistent. There are several patterns the locals use with a variety of retrieves, obviously imitating more than one insect. Late July and August are supposed to be the slow periods, but those who know the lake can enjoy explosive numbers of good fish during this "off" period. Some of the key areas include the springs in front of Staleys Springs Resort, the hatchery area and around the stream inlets feeding the lake.

Fishing from a boat or float tube is the best way to fish this lake. Wading is just about impossible in order to reach the openings among the weeds. Henry's has a two fish limit but you can catch and release numbers when you fish it right.

There are several resorts on the lake with Staleys Springs Resort, Wild Rose and Henry's Lake Lodge the most popular with fishermen and tourists. Staleys and Wild Rose have R.V. hook-ups and boat rentals and there is a county campground at the south end of the lake.

Weather patterns change quickly in the Rockies and wind is often a factor as it is on most lakes. Fall mornings can offer freezing temperatures so dress accordingly. And here's a tip: bring a camera so the folks at home will believe you.

This Yellowstone cutthroat is the mainstay of Henry's Lake with brook and Rainbow-/Cutthroat hybrids balancing out the fishery.

Fall is prime time on Henry's Lake as big hybrids move into shallow water along with large numbers of brook trout.

Henry's Lake Summary

SEASON: Opens Saturday of Memorial Day weekend and closes October 31. Legal fishing begins at 5 a.m. and ends at 9 p.m.

TROUT: Yellowstone cutthroat, cutthroat-rainbow hybrids, and brook trout. Hybrid cutthroat will average 19 inches, brook trout and cutthroat 16 inches although it varies over the years. All three species reach trophy size with cutthroats to 9 pounds, brookies to 6 pounds and hybrids to 18 pounds possible.

LAKE SIZE: Henry's Lake sprawls 6,500 acres with an average depth of 12 feet and the deepest hole about 25 feet.

RECOMMENDED FLY LINES: You can use a variety of lines here with a floating and intermediate line best for fishing shallow. I prefer to use the clear transparent line for flat water or to work depths of 6 to 12 feet and a type II or type III line to fish the deeper holes where many of the bigger hybrids lurk.

RECOMMENDED FLY PATTERNS: All the standard patterns will work here with Leeches and my Seal Bugger in olive and black good choices throughout the season. My Stillwater Nymph is an excellent choice during the summer damsels hatches and I prefer my Callibaetis Nymph, A.P. Emerger or Midge Larva when trout are working on or near the top.

BEST TIMES: There isn't a bad time for this lake. Fall is very good for the big hybrids and brook trout although you can take these fish all season.

Island Park Reservoir harbors some aggressive rainbow and lots of shallow shorelines for anglers to explore.

ISLAND PARK RESERVOIR
Idaho

Located only a short distance down stream from Henry's Lake, Island Park Reservoir has over 8,400 acres to cover and offers the stillwater angler an interesting challenge. The challenge lies in finding the trout. Although temperamental at times, this is an outstanding fishery. The lake has all the ingredients for supporting large numbers of big trout. It's shallow, full of weed beds and is rich in nutrients, all of which are part of the habitat that supports an abundance of aquatic insects and other food sources. These trout are rarely selective, but prolific insect hatches invite selective feeding. Most of the time, however, they feed pretty much opportunistically.

Rainbows, rainbow-cutthroat hybrids and Lahontan cutthroat comprise the fishery with kokanee providing added sport for the trollers. The average rainbow here runs 2 to 4 pounds with fish from 6 to 7 pounds. The hybrids are fast becoming the object of fly fishermen because of their ability to become double-digit fish, and they pull hard. The cutthroats, too, can put on the weight with fish to 8 pounds possible, but for a variety of reasons they are not the fly fishermen's darling.

During the spring trout tend to scatter because of the high water. Locating them becomes a priority if you want to play. The summer months, however, are another story. Most of the bigger fish will find refuge in cooler water or near the weedy areas when they feed.

This 7 pound rainbow is an example of the direction this fishery is going. (Dave Nolte photo)

Damselfly nymphs become the trout and the fly fishermen's main attraction. The hatch lasts well into early September so don't put those little greenies away before the action ends. Caddisfly and mayfly hatches can be heavy at times providing excellent opportunities for stillwater anglers on top as well as below the surface. Midges are fairly common and so are leeches and scuds.

When I fish Island Park, I prefer to search for trout using my Seal Bugger in both black and olive. That works well early and late in the day and during the spring and fall months. The rest of the time I'll pursue these fish with my Stillwater Nymph, my A.P. Emerger or Midge Larva. I like them all in olive when hatches bring the trout near the surface. Most of the popular nymphs work well here as well. The key once again is to locate these fish. If you want to pursue them on top, try a Parachute Adams, Pale Morning Dun, Light Cahill or Sedge Caddis to match the naturals.

Fly fishing really kicks in by mid-June and despite some on and off periods, the lake fishes well right into October.

Because of the numerous weed beds and shallow water areas, most of the lake's shoreline is productive, but not always at the same time. Knowledgeable fly fishermen fish the coves and points that surround the Fingers and Grizzly Springs areas near the West End Campground. Angling pressure often will tell you where the bite is.

Most of the bigger trout are pursued late spring into mid-June and again during the fall months of September and October. The area above McCrea Bridge and along the upper reaches of the reservoir can be hot for big fish following migrating kokanee in September. Where to fish in the summer months can be a coin flip. Where you end up should depend on water quality and water temperature.

The lake is best fished from a boat or float tube although wading can be effective when these trout move into the bays to feed.

This is Yellowstone country so be prepared for fast changing weather patterns and cold mornings.

An olive Seal Bugger triggered this response from a big rainbow near the Fingers area.

Camping is permitted, with R.V. and tent sites located along the west end of the lake. The West End Campground is reached by taking Green Canyon Road which begins at Harriman Bridge where Henry's Fork crosses Highway 220.

Island Park Reservoir Summary

SEASON: Open year-round.

TROUT: Rainbow, rainbow-cutthroat hybrids and Lahontan cutthroat with the average trout 2 to 4 pounds and the potential of bigger fish to 7 pounds possible.

LAKE SIZE: This one is big, 8,400 acres, but with lots of weedy shallow areas for stillwater anglers to explore.

RECOMMENDED FLY LINES: I like the intermediate or the clear transparent line that sinks about 2 inches per second. This allows you to fish the shallows effectively with the intermediate, or work the deeper areas with the faster sinker. Add the floating line if you want to work the surface or subsurface for emergers.

RECOMMENDED FLY PATTERNS: I like to explore with the Seal Bugger early or late in the day unless fish are working near the surface and certainly in the spring and fall months. My Stillwater Nymph is the best fly for me when the damsels are hatching. I'll also add my A.P. Emerger, Callibaetis Nymph or Midge Larva depending on what is happening on or near the surface.

BEST TIMES: June is good, but fall is best for the bigger fish. When these big trout move into the shallow weedy areas to feed during the warmer months, you should do the same.

Montana—Big brown trout are a large part of the fly fishing scene in Montana.

Chapter 5

Montana

The state of Montana needs no introduction to fly fishermen. Its waters have been the dream and focus of anglers regardless of their fishing preference. This is a state rich in fly fishing tradition, where hall of famers Dan Bailey and Joe Brooks along with other fly fishing legends helped establish their reputations. With so many blue ribbon streams and rivers to fish, there hasn't been a lot of ink regarding the many quality lakes and reservoirs found in this state.

I chose seven for my Top 50 but I could have easily picked several more. The Blackfoot Indian Reservation centered around Browning in the northern part of the state is loaded with quality stillwater lakes. Adjacent to Yellowstone country there are countless fisheries where rainbow, brown, brook and cutthroat trout abound. Many are good fly fishing lakes, others deep bodies of water that fish better with hardware where anglers can troll and dredge the depths.

Keep in mind if this is your destination this year, or in the future, you can bet it is on the agenda of others as well. Most of the better waters are crowded during the summer months, but there is lots of room early and late in the year. Spring is a good time for most lakes and fall is even better.

There is a lot to see and do in the Big Sky Country, which includes Yellowstone National Park in the southwest corner of the state with scenic wonders and historical points everywhere. My recommendation is to bring a float tube, fly rod and a camera when you go.

DUCK LAKE
Montana

This lake is the best big trout fishery found on the Blackfoot Indian Reservation. Located only a short distance from Glacier National Park in the northern edge of Montana, Duck Lake is in a class by itself. It offers the still-water angler the ultimate challenge for large rainbow and brown trout. I consider this lake as good for big trout as any I've fished in the United States. The lake holds large, broad-shouldered fish that will raise your blood pressure.

The fast growth rates stem from an enormous food base that quickly propels these trout into double digits. Scuds, caddisflies, mayflies, damselflies, leeches and forage fish are primarily responsible for their increased weight. The average rainbow, which is from the Eagle Lake strain in California, runs 3 to 6 pounds, but there are lots and lots of 8 to 10 pound fish with some to 17 pounds waiting to challenge your skills. The browns, although more elusive than the rainbows, run 2 to 5 pounds on average with fish from 10 to 15 pounds a distinct possibility. The challenge for most anglers is locating them.

Because there are no main tributaries into Duck Lake, each spring the big rainbows cruise the shallow bays and shorelines looking for areas to spawn. Although temperatures are on the cold side and the trout sluggish, these big fish can be caught. But it's not as much fun as when they have fully recovered from their spawning ordeal and are looking for food. In the fall it's the browns turn to cruise the shorelines. Most of the browns will begin to show in late September. It's never easy, but very rewarding when you land one of these monsters.

Sun up is prime time on Duck Lake when big rainbows and browns hunt in shallow water.

This 4½ pound brown was cruising shallow water when he took the author's Shiner Minnow. (Dave Nolte photo)

In July, Duck gets a big caddis hatch referred to as "Traveling Sedges" that bring these trout to the surface. You will find it challenging, exciting and exasperating all in one, but well worth the exuberance and the disappointment.

I prefer the intermediate line for both spring and fall fishing, especially when I use the Sillwater Nymph, a favorite for these big fish. The action never seems to slow for me with this fly. Used with short, slow pulls, it's deadly. But lots of patterns work. Besides, it's rarely the fault of the fly, only the guy using it. Consequently, the emphasis is on presentation as it should be.

Duck is shallow along its edges on the northern shoreline with some weedy areas at the west end. Fishing from shore, from a boat or in a tube will all work, but I prefer a tube or boat for the fall fishing. Access has become limited, but it's available at the launch ramp. When it's tough this lake has humbled many an angler, but when it's hitting on all cylinders it has made heroes out of beginners.

Duck is a far cry from what it used to be, yet still ranks as one the premier lakes in the country. If you doubt this claim, snuggle up to some of the old timers who have fished this lake over the years and just listen.

Stormy weather can move in quickly in the spring and fall along with extended windy periods. The sudden change in weather and temperature does a number on your hands so go prepared. (Wool, fingerless gloves are a good idea).

There is a Blackfoot run campground on the northern shore along with a launch ramp and small store where fishing licenses and some tackle can be purchased. Montana state fishing licenses are not required, but a tribal permit is. You must also have a boat or float tube permit to fish this lake.

Duck Lake is located just southeast of Glacier National Park on Highway 464.

This 8 pound rainbow was the victim of the author's stillwater system.

Duck Lake Summary

SEASON: Open year-round.

TROUT: Rainbow and brown with the average fish running 3 to 6 pounds with a lot of fish over 10. Browns and rainbows over 15 pounds have been landed in the past years.

LAKE SIZE: 450 acres with lots of shallow shorelines to explore.

RECOMMENDED FLY LINES: Duck is deep in many parts, but the best fly fishing isin shallow water so I use the clear transparent line that sinks about 2 inches per second and count down if I have to. For most of my spring or fall fishing I use the intermediate line. When Traveling Sedges are working in July, you will need a floating line.

RECOMMENDED FLY PATTERNS: I find my Stillwater Nymph the best fly for me when I fish in the spring and again in the fall. My olive bodied Seal Bugger is hard to beat for spring or fall fishing when most insects are absent.

BEST TIMES: Like most lakes, spring and fall are the best, particularly when these trout enter shallow water or cruise the shorelines. Right after ice out is another excellent time and will often last into mid-June. Summer fishing is good when the Traveling Sedges are present as fish over 10 pounds come up for these speed burners.

Spring fishing offers lots of surprises, but at Kipp Lake, big rainbows are not one of them.

<u>KIPP LAKE</u>
Montana

Although somewhat temperamental at times Kipp Lake is another of the excellent fly fishing lakes on the Blackfoot Indian Reservation. For a variety of reasons, the most obvious is its big trout, it makes my Top 50 list. But the one thing that draws anglers like myself to Kipp is the number of trophy trout this lake harbors.

Kipp is a relatively shallow lake. Its rich nutrients support healthy underwater vegetation and a host of scuds and aquatic insects that lend themselves to hungry and aggressive trout. These are Eagle Lake rainbows from California that will average 16 to 23 inches, but 8 to 10 pound trout are common.

Currently, the tribe has no fishing restrictions on the lake, but fly fishing is still the most consistent way of catching these fish. Insects abound here, with scuds active and available in April, May, September and October. Mayflies and damselflies are present from early summer into fall. However, the trout will search out leeches all summer long. I've enjoyed excellent action using my Stillwater Nymph and my Seal Bugger, but I've witnessed other anglers using a variety of patterns also catching fish.

The prime time on Kipp is early spring from ice out to early June, or until warm weather forces the trout into deeper water and out of the reach of shore anglers. You can tube or fish from a boat to reach these fish, but the bite is slower during the summer months. In the fall, fish will again move into shallow water providing some excellent action. The best area I've found to fish is off the low hanging cliffs on the opposite side of the lake from where you enter. A small dirt road circles the lake so access is not a problem. This opposite side has a gentle slope to the

A 7 pound Kipp Lake rainbow that ate a Stillwater Nymph in shallow water.

water and a rocky or gravel bottom. Scud populations are more exposed here and you will see trout cruising just off shore in search of these insects which are a staple in their diet. There are two small inlets, both of which will concentrate fish during the warmer months or when preparing to move up and spawn in the spring.

I find a soft tip rod and an intermediate line ideal for these big fish. You can cheat on tippet size and go to a faster tip rod for spring fishing when the wind can lift you off the ground when it gusts. However, more big fish are caught at this time than at any other time of year.

The lake is located about 6 miles east of the town of Browning off Highway 2. Then drive about a mile into the lake off a gravel road.

There is no improved camping here, but you can park it if you're self-contained. Lodging and supplies are available in Browning or in Cutbank about 25 miles east on Highway 2. As with all lakes on the reservation, a tribal fishing permit is necessary as well as a boat or float tube permit. However, no Montana state license is necessary when fishing on the reservation.

Kipp Lake Summary

SEASON: Open year-round.

TROUT: Rainbow with the average trout running 16 to 23 inches with 8 to 10 pound fish possible.

LAKE SIZE: This is a relatively small, shallow lake with lots of shallow shorelines to explore.

RECOMMENDED FLY LINES: An intermediate line fishes this lake best, but you may need the clear transparent line if the surface goes flat.

RECOMMENDED FLY PATTERNS: My Stillwater Nymph has been very good with my Seal Bugger a close second. Scud patterns will work here as will leech patterns when no surface activity is present.

BEST TIMES: This is a lake that is very good early in the season and slows considerably when the summer months arrive. Late season angling can be good when water temperatures cool.

Mission Lake is one of the more popular and productive lakes on the Blackfoot Indian Reservation.

MISSION LAKE
Montana

Mission Lake is a classic stillwater fishery located on the Blackfoot Indian Reservation in the northern part of Montana. Along with Duck Lake, Mission is a supreme big trout lake where trout under 16 inches are an absolute minority.

Mission is a long, shallow, extremely nutrient rich lake that is best fished from shore in the spring and from a boat or float tube in the summer and fall. Most of the rainbows in Mission average 3 to 6 pounds but 8 to 12 pound fish are not at all uncommon. Several strains of rainbow comprise the fishery with California's Eagle Lake strain the dominant adversary. As of this writing there were no forage fish for the trout to pursue, only insects, and lots of them. Scuds are a favorite of the trout and are available throughout the season along with caddisflies and Callibaetis mayflies.

Pattern selection isn't critical, but small nymphs like the Prince Nymph, Pheasant Tail, or Hares Ear, along with scud imitations, are seldom refused when these big trout are on the hunt. Leeches, my Seal Buggers, and Woolly Buggers are good patterns to fish, but don't command the attention that the smaller nymphs do. That is especially true in April and May when the bigger trout are in shallow water. I have never had a bad day on this lake using an intermediate line with a 12-foot leader, a 4x tippet and a size 10 Stillwater Nymph. In fact, it is so consistent I've rarely used other patterns. A slow, 4 inch pull-and-pause retrieve is quite effective and so is a slow, hand-twist retrieve. These fish concentrate, but keep in mind there is a lot of shoreline area that is unproductive. The key is locating where the trout are feeding and working those areas.

Because there is no tributary feeding Mission, the rainbows in the spring seek out the shallow gravel areas to drop their eggs. Without well aerated water to spawn in, the spawn never develops so the ritual ends up as a mere exercise for the trout. The cliffs along the northern shore are prime areas where the fish concentrate in the early season. The upper west end of the lake is shallow and weedy and is a favorite place to find these fish from late spring through the summer months.

Floating and intermediate lines both work and are especially effective when fishing the shallower regions of the lake.

Mission has earned its reputation as a trophy fishery, one well deserved if the increasing number of lake anglers is any indication. The lake is open all year, but freezes in late fall. The road leading into the lake, which is accessible off Highway 2, is a series of rolling hills with a clay base and can be a mess when it rains or snows. Wind is almost always present and necessary for a consistent bite. It is never a guarantee but you will struggle without it.

Camping is permitted, but since this is on Blackfoot land, permits are required for anything you plan on doing on the reservation. Lodging and fishing permits along with supplies are available in Cutbank and Browning about 15 miles from the turn-off to the lake in either direction.

Large trout like Dave Nolte's 10 pounder are a big part of the expectations when fishing Mission Lake.

Spring fishing at Mission lake is a magical time providing anglers with size and spectacular color.

Mission Lake Summary

SEASON: Open year-round.

TROUT: Rainbow with a healthy average trout running 3 to 6 pounds. There are some 8 to 10 pound lunkers around although this lake had to start over in 1998 so the average trout may be smaller.

LAKE SIZE: Mission is a long lake, shallow along its shoreline edges, but it doesn't fish well from a floating device except at the upper west end where the weed beds harbor some big trout.

RECOMMENDED FLY LINES: This shallow lake is an intermediate or floating line challenge.

RECOMMENDED FLY PATTERNS: My best pattern here is the Stillwater Nymph. Lots of standard flies work. Trout are not picky about your choice of pattern, but how and where you present it.

BEST TIMES: The lake is at its best right after ice out through early June and again late in the season.

MITTEN LAKE
Montana

Considered one of the better fly fishing lakes by many anglers who fish the Blackfoot Reservation regularly, Mitton Lake is also one of the most pristine of all the reservation lakes. It has it all: plenty of shallow water, big, full-shoul-dered rainbows, lush green meadows, stands of pines and aspens, and of course, the solitude that makes it all worth while.

Mitten is about 170 acres averaging about 7 feet in depth. The rainbows that call Mitten home aren't huge, but have the reputation of being powerful fighters, capable of tearing the rod from your grip if you are not paying atten-tion. Most of these rainbows run 3 to 4 pounds with some reaching 5 to 6 pounds on occasion. Regular plants of 20,000 to 35,000 Arlee rainbows are stocked each year to maintain the fishery.

My Stillwater Nymph was a big fly for me when I last fished this lake, but I think most good suggestive nymphs will work. Early and late in the day, I used my olive Seal Bugger with good results. You can catch bigger fish on many of the other reservation lakes, but pound for pound, none fight any bet-ter than these rainbows.

I found the clear and intermediate lines good choices for fishing Mitten. On flat days, without a breeze to ripple the water, dry fly action can be good. When the bite is happening here, most lines can be used, but I don't believe they will produce as well as the clear or intermediate line.

Mitten Lake rainbows don't get huge, but run 2-5 pounds like this 4 pounder.

There really isn't a bad time to fish Mitten. Most areas of the lake contain fish with the rainbows cruising shallow water early and late in the day. Weather can be harsh up here especially in spring and late fall. When water temperatures are on the chilly side, as they usually are early and late in the season, use slow retrieves. You will find they will work better and are more consistent.

Leaders up to 12 feet with 4x tippets are good choices here and can be altered depending on conditions and the fly you are using.

There are no facilities at Mitten and you must have a tribal fishing license to fish here, as well as other waters on the reservation. Be sure to check the regulations for times, seasons, use permits for float tubes, life vests, etc.

If you are going to fish the reservation, this one is a must.

Mitten Lake Summary

SEASON: Open year-round.

TROUT: Rainbow with the average trout running 3 to 4 pounds with some bigger. These are broad shouldered fish with plenty of gusto.

LAKE SIZE: Mitten has about 170 acres averaging about 7 feet in depth.

RECOMMENDED FLY LINES: All I have used here is the intermediate slow sinking line, but I imagine other successful anglers have used a medley of lines that have worked.

RECOMMENDED FLY PATTERNS: My Stillwater Nymph was excellent during the spring and again in the fall. I've also used my olive Seal Bugger with good results.

BEST TIMES: There isn't a bad time for this lake, but I think the action is a bit more consistent in spring and fall.

CLARK CANYON RESERVOIR
Montana

Fast becoming one of the mega trout fisheries in Montana, Clark Canyon Reservoir is a lake attracting increasing numbers of stillwater anglers each year seeking her double-digit trout.

Only a few years ago, rainbows averaged 1 to 2 pounds and the browns were a respectable 3 to 4 pounds. Today, the 'bows are running 2 to 5 pounds with 6 to 10 pounders fairly common. Meanwhile the browns, although fewer in number, average 3 to 6 pounds with fish to 16 pounds possible. Part of the reason for the large fish is the presence of a freshwater ling called a "burbot." The fry of these fish allow the browns and rainbows to add weight throughout the year.

Leeches, Woolly Buggers, my Seal Buggers along with assorted minnow imitations, are excellent choices during the spring and fall months when other food sources are less available. During the summer months, the lake supports excellent hatches of aquatic insects to the delight of fly fishermen. Callibaetis mayflies begin appearing as early as May and continue through most of July when damselflies become the prime target of the trout. The Callibaetis reappear in the fall and can be found on the water into October, weather permitting. Midges are available all year, although the trout seem to prefer other insects during the summer when their numbers are intense.

Clark Canyon Reservoir and the mouth of Red Rock Creek form a gathering place for big trout both spring and fall.

Because the rainbows and browns sometimes vary on food choices, floating, intermediate, the clear transparent, along with uniform sink lines are necessary to match the changing conditions. Standard leader lengths of 10 to 12 feet with tippets to match your patterns will work here. In the spring heavier tippets to 3x are a wise choice when working minnow patterns.

In a normal water year the lake covers 6,500 surface acres. The Beaverhead River, famous for its outstanding brown trout fishery, drains the reservoir which was formed by damming the Red Rock River and Horse Prairie Creek.

The most popular area for fly fishermen is the upper end around the weed beds where the greatest amount of food is available. During high water years, the inlet area of Red Rock River is especially good. And during low water periods, the Horse Prairie Creek area is usually productive. The lower end of the reservoir near the dam is better suited for trollers and bank fishermen.

The reservoir is open year-round, but it freezes in the winter. The best fly fishing occurs after ice out, again in the early summer when hatches begin in earnest, then again in the fall when most anglers have long departed. My personal preference is in the fall when the browns are more concentrated. I've enjoyed spectacular fishing using my Seal Buggers with both black and olive bodies early and late in the day. My Stillwater Nymph is a fish catcher in the late mornings and into the early afternoons when the damsels are active. When the adult Callibaetis and midges appear, typically in the late morning and early afternoon hours, my Tan Callibaetis or olive A.P. Emerger are also productive patterns.

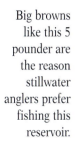

Big browns like this 5 pounder are the reason stillwater anglers prefer fishing this reservoir.

Fishing from boats and float tubes is effective, but wading is not. Wind is always a factor on lakes and Clark Canyon is no exception. Spring and fall weather can change quickly so dress for cold. There is camping on the lake but motels, supplies and other amenities are available in Dillon.

Clark Canyon is located 20 miles southwest of Dillon just off I-15.

Clark Canyon Reservoir Summary

SEASON: Open year-round.

TROUT: Rainbow and brown trout with the average rainbow running 2 to 5 pounds and 'bows to 10 pounds possible. The browns will average 3 to 6 pounds with some reaching 15 and 16 pounds.

LAKE SIZE: This reservoir is big with 6,500 acres when full. There are lots of shallow shoal areas to fish, but shore fishing is not very productive.

RECOMMENDED FLY LINES: I find the intermediate or the clear transparent lines best for this lake. You may find a need for a uniform sink II if you want to probe the bottom areas and possibly a floater if you prefer to chase them on top.

RECOMMENDED FLY PATTERNS: My Seal Bugger in olive or black, especially the latter during low light conditions have been very consistent. I'll use my Callibaetis Nymph or A.P. Emerger in tan or olive when adult Callibaetis or midges appear.

BEST TIMES: The reservoir is at its best spring and fall with summer fly fishing good if you fish the top few feet with small nymphs. It is hard to beat the fall fishing for big rainbows and browns where Red Rock Creek enters.

Early morning, a heavy chop and a black Seal Bugger were a deadly combination for Dave Freel.

Georgetown
Lake is
excellent
fly fishing
water if you
have an
intermediate
line—a long
day if you
don't.

GEORGETOWN LAKE
Montana

At first glance Georgetown Lake looks better suited for water skiers than fly fishermen, but the saying "looks can be deceiving" is applicable here. I first heard about Georgetown Lake when I was a young boy. I was fishing for bluegill next to two elderly gentlemen on a lake in California. They were discussing the large rainbows in Georgetown Lake and even though I was just a kid, their stories and conversations fascinated me. Several years later I learned to pursue trout with a fly rod, but it wasn't until 1972 that I first fished this mystifying lake. Wow!!

At the time, the rainbows were big, some to 6 pounds, and plentiful. But the lake went through a period where fishing declined, common with a lot of stillwater fisheries. But, after being down for a few years, the big fish are back and in large numbers again. The rainbows average 17 to 22 inches (about 2 to 4 pounds). Bigger rainbows from 5 to 6 pounds are still fairly common with fish from 8 to 10 pounds possible, although badly outnumbered by their smaller cousins. Brookies and a few bull trout share space with the rainbows and on occasion, a bull trout over 10 pounds is landed. Those old guys would be proud.

Georgetown is a rich lake, somewhat shallow with lots of weed beds and has a good mixture of aquatic insects. Midges and caddisflies dominate the scene, but it is the damselfly and dragonfly nymphs that excite these big trout. Try using a Midge Pupa in sizes 16 or 18, Sedge Caddis in sizes 14 or 16, olive damsel nymphs and Carey Specials in sizes 8 to 12 along with Woolly Buggers and Leeches in sizes 6 or 8 to match the food sources these trout look for. The bull trout feed mainly on forage fish or small trout and are most vulnerable in the spring and fall. Even though their numbers are few, when you hook one of these trout you know it's going to be big.

My Stillwater Nymph has been far and away the best pattern for me on this lake from June through September matching the damsel nymph. For the midges or mayflies, I use my tan or olive Callibaetis Nymph, olive or black Midge Larva or Black Diamond. The olive bodied Seal Bugger with a burnt-orange hackle is good for an early and late season pursuit of rainbow or bull trout.

Most fly lines will work, but on this shallow, weedy lake, the intermediate is best. If you find fish feeding in clear, shallow water or when the water is without ripples, the transparent line is a better choice.

The best times for me have been late spring and fall for the largest fish, but numbers of 16 to 20 inch rainbow are easier to come by beginning in the summer months and lasting into September when hatches of damsels are in full-swing.

Georgetown has about 3,000 acres and is located 20 miles west of the mining town of Anaconda. The season begins the third Saturday in May and closes March 31.

This chunky 4$\frac{1}{2}$ pound rainbow is the result of the damsel nymph action that occurs each summer on Georgetown Lake.

The winds can blow here and there is never a guarantee on what the weather might do so be prepared for the worst.

Access is no problem and with plenty of shoreline available, wading is very effective. Fishing from float tubes and small boats is just as good especially around the shallow weedy areas. I've experienced my best fishing off the shallow west side of the lake. This is a popular area and for good reason: People catch a lot of big fish here. I believe it's most consistent in the spring and fall or early or late in the day. The weed beds will claim some fish so adjust your tippets to the patterns and conditions at the time. Pull your flies slow and hang on. You won't be disappointed.

There is camping on the lake, but all other amenities are in Anaconda.

Georgetown Lake Summary

SEASON: Georgetown Lake is open the third Saturday in May through March 31. The south and east shorelines are closed to fishing from shore or within 100 yards of shore from April 1 until July 1. The closed area extends from a point 200 yards west of Denton's Point Marina (along the shore, including all of Stuart Mill Bay and its entirety) to a point 200 yards north from the mouth of the North Flint Creek.

TROUT: Rainbow, brook and bull trout make up the fishery. Rainbows average 17 to 22 inches. Fish to 5 or 6 pounds are common with a few reaching 10 pounds. The bull trout, although fewer in numbers, are big with some well over 10 pounds.

LAKE SIZE: Georgetown has about 3,000 acres with lots of shallow shoreline access.

RECOMMENDED FLY LINES: The intermediate or the clear transparent line should match most conditions you will encounter. Other lines will work but not as well as these two.

RECOMMENDED FLY PATTERNS: The best fishing here is with damsel nymphs, although many standard aquatic nymph patterns will work. I prefer my Stillwater Nymph in size 10 fished just under the surface. If your pursuit is for the bull trout, think minnow. My Seal Bugger or Woolly Bugger patterns are also good choices.

BEST TIMES: The spring and fall are fairly consistent for big trout. However, it is difficult to not take advantage of the damsel hatch that starts in late June and lasts into early September.

Dave Freel
plays a nice
Hebgen Lake
brown.

HEBGEN LAKE
Montana

Hebgen Lake interrupts the upper Madison River some 20 miles from Yellowstone National Park. Unlike most other lakes, Hebgen offer anglers some unusual action for big trout otherwise known as "gulper fishing". The term originated here on Hebgen and is used to describe the browns and rainbows cruising indiscriminately while intercepting adult midges, caddisflies or mayflies lying on the surface.

Most of this action takes place in the upper regions of the lake where the river inlets and shallow bays offer trout food, cover and cooler water. Hebgen has rainbows from 14 inches to 5 pounds, browns from 2 to 5 pounds, and cutthroats from 1 to 3 pounds although their numbers are declining. There are bigger fish available and occasionally fish from 6 to 8 pounds are landed.

Prime time for "gulper activity" starts in July and runs through August when hatches are intense and frequent. Matching patterns include Griffith's Gnat in sizes 16 to 20 for the midges, Parachute Adams in size 14 to 18 for the mayflies, and the Elk Hair Caddis in sizes 12 to 16 to match the caddis.

Standard nymphs, along with Leeches and Woolly Buggers, are also excellent choices if you fish below the surface. Most fly fishermen refer to Hebgen's "gulper action" as a dry fly affair, but the last time I fished this lake I used nymphs just below the surface. I rigged up with my intermediate sink tip, a 15 foot leader with a 5x tippet and used my tan and olive Callibaetis nymphs. I hooked 16 fish, ten rainbows and six browns that ranged from 16 to 22 inches. I broke off one that pulled like a semi-truck bound for Yellowstone. After a late morning cookie break, I switched to my Stillwater Nymph just to see if they would eat it. Although I spotted only a handful of damsels on the water, I used a slow, hand-twist retrieve, and the cruising fish that I didn't spook with my cast took it aggressively. About three hours later I had landed 13 fish, including one 'bow about 6½ pounds, and called it a day.

If you fish the dry fly, remember to lead the fish you're casting to about two to three feet and let the fly sit. If you don't move them off their cruising path, anticipate the take if you are matched up with what they are eating.

Most of the action can be matched using floating lines and long leaders measuring 12 to 18 feet with tippets to match the pattern. Nymphing can be done with either the floater or an intermediate line with the latter a better choice if the wind picks up.

Fly fishermen who have experienced this action know how addictive it can be. Most use float tubes, but boats are just as effective and can provide more mobility if you want to move a lot. Wading is out of the question as you can't reach working fish as easily.

A Hebgen Lake brown that chose a Stillwater Nymph over an array of naturals on the surface.

Hebgen is actually a reservoir interrupting the Madison River before dumping into Quake Lake, another outstanding but temperamental fishery located a few miles down stream. Hebgen is open all year, but is best for fly fishermen starting in July through September.

There are a number of campgrounds in the area and with West Yellowstone only 30 minutes away, motel accommodations, restaurants, and supplies are all convenient. All the fly shops offer guides and up to date fishing information.

If you haven't fished Hebgen yet, do it. I think you'll be back.

Hebgen Lake, where the Madison River enters, offers excellent fall action for rainbows and browns

This rainbow was gulping (surface feeding on the move) when he took Denny's Tan Callibaetis nymph fished on an intermediate line and a hand twist retrieve.

Hebgen Lake Summary

SEASON: Open year-round.

TROUT: Rainbow, brown and cutthroat. Rainbows will average 14 to 20 inches with fish to 5 pounds and an occasional lunker taken. The browns run 2 to 5 pounds, but 8 to 10 pound fish are landed by trollers from time to time in the spring or fall. Cutthroats are declining with the average fish 1 to 3 pounds.

LAKE SIZE: This is a big lake with an average depth 8 to 12 feet. The upper reaches are somewhat shallow, but fishable.

RECOMMENDED FLY LINES: Most of the gulper action is done with floating lines that can fish the dries on top or the sub-surface action just below. I prefer using an intermediate or intermediate sink tip that will keep my fly in the top foot. You can use faster sinking lines with leeches, Woolly Buggers or my Seal Bugger, but the fun here is on or near the surface.

RECOMMENDED FLY PATTERNS: Griffith's Gnat, Parachute Adams or an Elk Hair Caddis will usually cover the surface action. I do well with my tan or olive Callibaetis and Stillwater nymphs in size 12. Remember, you need to keep them in the top 6 inches for consistent action.

BEST TIMES: This is summer action with late July through August being excellent.

Nevada—Mike Dehart in pursuit of big rainbow from a remote high desert Nevada lake.

Chapter 6

Nevada

Known more for its painted deserts, sage brush, jack rabbits and gambling casinos, Nevada remains a sleeper when it comes to still-water fly fishing.

Its semi arid climate, alkaline flats and high desert atmosphere are well defined and perfect for growing big trout. If there is a negative here, it's a lack of water holes where fly fishermen and other anglers can play.

Most stillwater anglers know of or have fished for the big Lahanton cut-throat in Pyramid Lake near Reno, or for the big rainbows in Ruby Marshes or Wildhorse Reservoir near Elko. But, few have tested the big rainbows in the nutrient rich water of Knott Creek or Sheep Creek reservoirs. Conversely, lakes such as Tahoe, Walker, Mead and Rye Patch are big bodies of water offering quality angling for a wide variety of species, but don't offer fly fishermen much opportunity. Still, there are numerous top notch fly fishing waters tucked in and around out-of-the-way places all over the state. It would be easy to mention a half dozen other small lakes that support some big trout, but their fragile habitat would place their future in jeopardy. With such a low number of licensed anglers in the Silver State, crowding is seldom a problem. So, get out, snoop around, locate some of these hidden gems and enjoy.

Knott Creek Reservoir with its unique rock formations offers anglers excellent fly fishing opportunities.

KNOTT CREEK RESERVOIR
Nevada

I first heard of Knott Creek Reservoir from a Nevada Fish and Game warden in 1988 while I was giving a slide show on stillwater fly fishing. He spoke of big rainbows and tiger trout, a cross between a brown and brook trout. When he mentioned that the lake didn't get a lot of pressure and fly fished very well, he had my attention. In July of 1998, I fished Knott Creek for the first time with two friends from Klamath Falls, Oregon, Vic Downer and J.D. Bell. Wow, what a gorgeous place!

Knott Creek is a relatively small reservoir stretching out to about 70 surface acres. Located in a remote area of Humbolt County, it is close to the eastern boundary of Sheldon National Wildlife Refuge. From stories I've heard, the real big rainbows are a lot fewer now as are the tiger trout, which I have yet to land. Knott Creek suffers from the same problem all reservoirs have difficulty with: up and down water levels. Food sources are greatly diminished while established weed beds are left high and dry to bake in the sun after water levels drop. Although the numbers of 6 to 10 pound 'bows are fewer now, there still are healthy numbers of 16 to 22 inch rainbows to test your skills. As for the tiger trout, there are still some 15 to 20 inch and a few bigger ones left but my understanding is they will not be stocked any more in future years.

Knott Creek is much more than just a good fishery. Its tranquil setting and solitude are unique, not to mention how remote this place is. The granite rock formations and wildlife are all part of what makes this place special.

This is really a fly fisherman's lake, shallow, nutrient-rich, light algae, weedy shorelines with intense hatches of aquatic insects. During June and July, damselflies and Callibaetis mayflies dominate the scene, commanding the trouts' attention. Midges are available all season while scuds are there for the taking throughout the spring and fall. I found my Stillwater and Peacock Callibaetis Nymphs in size 10 absolutely deadly when I fished it in July of 1998 and again in 1999. Some nice fish were also taken on my olive Seal Bugger early and late in the day.

Because the shorelines are shallow and weedy, I used my intermediate line and Cortland's new clear 15 foot sink tip with outstanding results. I see no reason to use fast sinkers here unless you want to coax non-feeders. There are dry fly opportunities all season especially with adult damsels fished around the weed bed edges.

A Knott Creek rainbow displays his leaping ability.

When fishing nymphs, I stayed with a 12 foot leader and a 4x tippet. You can play with this one depending on present conditions. If you challenge these trout with the dry fly, you may need to extend the leader a bit and go to 5x on the tippets.

Weeds can be a problem in low water years or during the fall months after drawdowns occur. Lake temperatures will rise during the summer months, which often have a negative affect taking the trout off their bite as the pH level goes up.

Normally the best fishing at Knott Creek occurs in the spring as soon as you can get in and again in the late fall. You can take fish all summer, but it is not as good.

Nevada regulations call for catch and release only from the second Saturday in May through the second Friday in June. Then from the second Saturday in June through November 15, the limit is one trout 18 inches or more. Only lures or flies with single barbless hooks are allowed and if you are so inclined, night fishing is legal.

Boats and motors are allowed, but a flat wake and a 5 m.p.h. speed limit must be observed. Most of the fishing here is done from float tubes, pontoon boats or by wading. The lake can be easily waded and is a good idea when the damsels are hatching. If you choose to bring a boat, you must hand launch it as there are no boat ramps. Besides, I'm not sure it is a good idea to pull a trailer over that goat trail they call a road. I was informed they have improved the road going in so you may want to inquire.

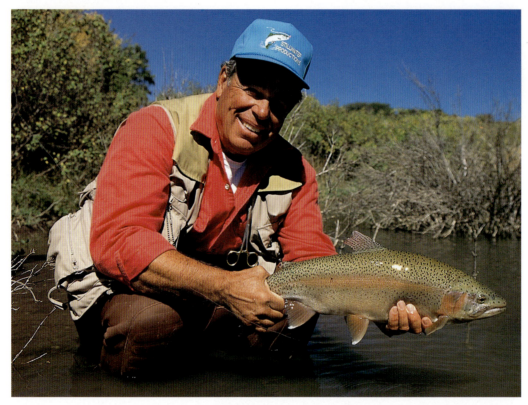

The author with a 5 pound rainbow while wading the shoreline with his Olive Peacock Callibaetis Nymph and an intermediate line. (Dave Nolte photo)

Camping is unimproved, no restrooms or water so plan on bringing in all the necessary supplies including tarps for protection against sudden storms.

To find Knott Creek Reservoir, take Highway 140 west from Denio Junction, Nevada, go about 13 miles and turn left at the sign that says Knott Creek. Drive 9 miles and turn left at the next sign. Drive another 4 miles until you reach the Onion Lake fork. This becomes a 4-wheel drive portion if it rains or the road is muddy. Go another 4 miles to Roy Rogers Rock as they call it. It's downhill from here and about another mile to the lake. If you own stock, ride in. If you have deep pockets, hire a helicopter. If you drive, bring a mechanic. Seriously, it is a rough road in a few places and not recommended for passenger cars or vehicles without 4-wheel drive.

On the other side of the coin, this road helps to keep the pressure down and gives the trout a break.

Knott Creek Reservoir Summary

SEASON: Nevada regulations call for catch and release only from the second Saturday in May through the second Friday in June. Then from the second Saturday in June through November 15, the limit is one trout 18 inches or more. Only lures and flies with single barbless hooks are allowed with night fishing legal.

TROUT: Rainbow make up the bulk of this fishery with the average 'bow running 16 to 22 inches with fish to 8 to 10 pounds possible, but their numbers are few. Tiger trout, a cross between a brown and brook, are also present but their numbers are declining as well.

LAKE SIZE: This is a shallow lake, about 100 to 110 acres when full. Lots of shallow shorelines areas to fish which cater more to wading or float tube fishing.

RECOMMENDED FLY LINES: With so much shallow water to explore, a floating or intermediate sinking line is all you need.

RECOMMENDED FLIES: I found my Stillwater Nymph, Callibaetis Nymph, Peacock Callibaetis Nymph or Seal Buggers match up well throughout the season.

BEST TIME: Spring and fall are best here but you can take fish all season. Concentrate on the shallow shorelines spring and fall and mid-depths when the heat arrives.

The line-up
standing on
their ladders
near Sutcliffe
on Pyramid
Lake.

PYRAMID LAKE
Nevada

Located 30 miles north of Reno on the Paiute Indian Reservation, Pyramid Lake is home to the largest cutthroat trout anglers will ever challenge. Surrounded by sage brush flats and soft rolling hills, Pyramid is a high alkaline desert lake where the Lahontan cutthroat has learned to adapt over the centuries. The world record cutthroat, a 41 pound monster, was landed in 1925 by John Skimmerhorn and researchers believe bigger fish were taken but not recorded. Today, anglers from around the West come to stalk these giants that average 4 to 5 pounds, but double-digit fish are very common. Before the turn of the century 20-pound cutts were numerous, but in past years only a few exceeding 20 pounds have been landed.

Pyramid is really a winter fishery with the best angling extending from November through mid-April at which time the trout leave the shallows and head for deeper, cooler water. This is a huge lake with over 70 miles of shoreline and depths extending to almost 300 feet. But fly fishermen are most successful wading rather than fishing from a boat or float tube. When feeding, these big cutthroats cruise the shallow sandy bays and rocky drop-offs in search of their main forage fish, the tui chub.

When I first fished the lake in 1968, the Woolly Worm in sizes 4 to 8 was the best and most consistent fly and it still is today, although I believe the Woolly Bugger and my Seal Bugger are just as good. Black, dark olive, purple and pink are the favorite colors, but everybody experiments when fishing is slow. Most anglers favor a 9 to 9 1/2 foot graphite rod with 7-8 weight shooting heads that are fast sinkers. The idea is to cast as far as possible, get the fly to the bottom quickly and then begin a slow, steady retrieve. The cutthroats here seldom take it far out, but prefer to follow the fly in then eat it as it is lifted off the bottom and heads to the surface. The strikes are rarely vicious, but solid enough to stop your fly before the fish turns and heads for deep water.

Leaders can be relatively short, 6 to 7 feet in order to sink the fly quickly. Because they rarely prove leader shy, tippets of 2x or 3x are enough to hold them. These big cutthroats are seldom selective and don't require a lot of finesse, but you do need time and patience to learn the fishery. Windy days or choppy water tend to spur the cutthroats into feeding. Having said that, there will be some days where there are endless cycles of casting and retrieving without much interest from the fish. I don't believe it's so much a matter of refusing your offering as it is staying out beyond casting distance until the wind stirs things up and brings them into shallow water to feed.

Although these giants may show anywhere, the most consistent and popular areas for fly fishermen are the Nets, Warrior Point, Block House and Pelican Point on the west shore and Dago Bay near the mouth of the Truckee River at the south end. The bite can vary some years depending on the amount of snow pack in the Sierras, which at times will result in runoff from the Truckee River into the lake. It may sound strange, but I've enjoyed my best fishing following light winters.

Because this is a winter fishery, hands and feet are going to be tested so dress accordingly. Neoprene waders are a must along with warm wool hats and hooded jackets that will withstand the wind, cold and moisture.

A Paiute Indian permit is necessary to fish on the reservation and can be purchased along with a boat or camping permit in the small community of

Most of the cutthroat landed at Pyramid by fly fishermen are taken by wading.

Sutcliffe located right on the lake. Extreme caution should be used if you fish from a boat; it gets treacherous and very rough quickly.

Up to date fishing information is available from The Reno Fly Shop in Reno (775) 825-3474.

To get there, take Highway 445 out of Sparks and that will take you to this desert giant.

Pyramid Lake Summary

SEASON: Open year-round.

TROUT: Lahontan cutthroat with the average fish running 4 to 6 pounds with 10 pound fish common and monsters to 18 to 20 possible.

LAKE SIZE: This is a huge fishery to cover, but the best action is always close to shore especially when the wind is blowing in creating ripples or breakers for the cutthroat to feed under.

RECOMMENDED FLY LINES: A shooting head that sinks fast to get to the sandy bottom is key to catching these big trout, but you can use any full sink line as well.

RECOMMENDED FLY PATTERNS: Woolly Worms, Woolly Buggers and some leech patterns make up the majority of what anglers use. I found my Seal Bugger a consistent pattern. Color is often more important than pattern with black, olive, purple and pink hot colors.

BEST TIMES: This is a winter fishery that begins in November and runs through April. These trout will go deep during the warmer months and will be out of reach of fly anglers.

Sheep Creek Reservoir supports good numbers of big trout like this 5 pound rainbow.

SHEEP CREEK RESERVOIR
Nevada

Sheep Creek Reservoir is a little known and lightly fished lake on the Duck Valley Indian Reservation near the Idaho border in Northern Nevada. It is also home to some very large trout.

Surrounded by sage brush, this high desert lake is shallow, with the deepest area just over 20 feet, and it has about 1,000 surface acres for fly fishermen to explore. The reservoir is rich in aquatic vegetation and has sufficient insects to keep both the fish and anglers happy. Callibaetis mayflies, midges and scuds along with damselfly and dragonfly nymphs and small forage fish supply the calories for these big trout.

The lake's main challenge for fishermen are Kamloops and Eagle Lake strains of rainbows that often exceed 10 pounds with fish to 15 pounds already landed. These trout are active early mornings and late evenings just off shore near the weedy areas. Daytime temperatures keep most of the trout deep and somewhat inactive.

Insect hatches will bring trout to the surface allowing you to experience some hot action with patterns that correctly match the hatch. But don't expect the fish to be the larger members of the lake's trout population. The sought after fish are usually taken by anglers using streamer or leech patterns and Woolly Buggers in black, brown and olive. My favorites here are my black or olive Seal Buggers with a burnt orange hackle in size 8 for spring and fall fishing, my Stillwater Nymph, a peacock or gray Callibaetis

Nymph, or my olive Midge Larva all in sizes 10 or 12 when surface activity occurs. I prefer a long, slow pull with the Seal Buggers and a slow, hand-twist retrieve with the nymphs.

A floating or intermediate line is all you need to be successful here. Heavier tippets of 3x and 4x can be used in the spring months when nutrient colored water helps protect presentation mistakes. I still prefer a 12 foot leader, but you can get by with a 9 footer if you prefer.

Sheep Creek fishes best during early spring and again in the fall, particularly early mornings and late evenings. The bite slows during the summer months as day time temperatures will run in the 90s and above. Access is easy and you can do well wading, tubing or fishing from a boat. The reservoir can be reached off state Highway 11, about 12 miles from Owyhee after which you drive on an improved dirt road to the lake.

The Shoshone-Paiute Tribe charges for camping and fishing. Permits for fishing run $5.00 a day, $9.00 for two days and $4.00 every day thereafter. Camping is $3.00 a day.

Like so many off-road lakes, it's rarely crowded, but the size of the trout makes you wonder why. I suggest bringing a camera.

This 7 pound buck was taken in shallow water on a black body, burgundy hackled Seal Bugger and an intermediate line.

Sheep Creek Reservoir Summary

SEASON: Open year-round.

TROUT: Kamloops and Eagle Lake rainbow are the object of angler attention (and affection) here. The average trout runs 2 to 4 pounds but trout to 10 pounds are certainly possible.

LAKE SIZE: Sheep Creek is a shallow lake with about 1,000 surface acres when full.

RECOMMENDED FLY LINES: I found the intermediate line works best when I fished it and if you fish here in the spring or fall, you can use a floater when these trout cruise shallow water.

RECOMMENDED FLY PATTERNS: I found my olive and black Seal Buggers were consistent in the spring while my Stillwater, peacock or grey Callibaetis Nymphs. My Midge Larva worked best when the hatches begin late spring and into the summer months, although these hatches may be brief during these hot periods. The bite is usually consistent early in the day,

BEST TIMES: Spring and fall are especially good following ice out, but early and late evenings can be good during the summer months.

WILDHORSE RESERVOIR
Nevada

If high desert lakes turn you on, this one's for you. Wildhorse is a reservoir I've known about for over thirty-five years and whenever it becomes a topic of conversation, it's measured in terms of pounds, not numbers.

Like all reservoirs, it has had its ups and downs but always bounces back from the off years.

The lake is located about 65 miles north of Elko, just off Highway 225. Like most alkaline desert lakes, Wildhorse has big trout and a variety of species to choose from. Rainbows to 8 pounds have been taken and I know there are bigger ones available. Browns to 13 pounds are not uncommon and cutthroats, along with the rare hybrid species tiger trout, have been landed over 10 pounds. The lake also contains smallmouth bass, white crappie and channel catfish.

Pyramid Lake may be Nevada's main attraction, but I think Wildhorse produces more pounds of trout per angler. This happens for two reasons: less water to cover and a more consistent bite. However, the lake is not small. It's about 5 miles long and about 1 1/2 miles wide at the widest point. The maximum depth when full is about 75 feet but there is lots of shallow water for good fly fishing.

As with all high nutrient lakes, Wildhorse is blessed with multiple hatches of aquatic insects, weedy areas and algae for cover. Callibaetis mayflies, midges, damsel and dragonfly nymphs along with small forage fish provide plenty of opportunity for these trout to grow big and fast.

These big trout are not familiar with Seal Buggers, nor are they accustomed to their life-like movements or their suggestive appearance, yet when I last fished there, they were consumed with alarming regularity.

Anglers using olive damsel nymphs, Hares Ears, Zug Bugs and Pheasant Tails in sizes 12 to 14 will find plenty of action. Most of the bigger trout are taken on streamer patterns like the Zonker, Matukas and white marabous in sizes 6 or 8. Woolly Buggers and Leeches will prove just as good in the same sizes. I've always done well using my Seal Bugger in black or olive, along with my Stillwater Nymph in size 10.

Wildhorse is at its best early and late in the day especially during late March if it's not still frozen. April, May and early June are excellent months to fish and the same can be said from October through November. The summer months will be slower for trout; however, it's reasonable to expect to catch trout early mornings when they come into shallow water to feed.

The trout are scattered but a popular area for fly fishermen is along the southern shore where there is an abundance of weeds, shallow bays and channels. Wind can be a factor here so go prepared.

Wildhorse State Park offers a boat ramp and campground with all the amenities for campers. The Bureau of Land Management also provides a campground near the dam area.

Wildhorse Reservoir Summary

SEASON: Open year-round.

TROUT: Rainbow, brown, cutthroat and hybrid tiger trout offer plenty of variety. The browns, cutts and tigers are few in number allowing the more plentiful rainbows to dominate the catch of anglers. Trout to 10 pounds are always a possibility with lots of 3 to 5 pound fish available.

LAKE SIZE: Although this is a big lake, there are plenty of shallow shoreline areas for anglers to explore.

RECOMMENDED FLY LINES: I find the intermediate the best line for me but if you want to probe the depths try the faster sinking lines. Deep water is usually not that productive except in the spring when the warmer water is near the bottom.

RECOMMENDED FLY PATTERNS: Lots of flies work here. My Seal Bugger in olive or black is very good in the spring and fall. When hatches begin in late spring, my Stillwater Nymph and my A.P. Emerger are very consistent.

BEST TIMES: Spring and fall are best and especially after ice out. Summer months are generally slow as these trout tend to be sluggish then, but early and late evenings can be good.

New Mexico—Dave Nolte with a 6 pound Jicarilla Apache Reservation rainbow.

Chapter 7

New Mexico

With much of its territory situated in a high desert setting, New Mexico has some gorgeous country and exceptional fly fishing spots to explore. Much of the state is flat and dry, but in the northern reaches it's mountainous, rugged and quite scenic. Sadly, I haven't spent enough time in this state to qualify all of its better stillwater fisheries. Time will change that.

Many of the lakes and reservoirs are on private ranches or Indian reservations. There are some exceptions, but you will need to cover some ground to fish them.

The Jicarilla Apache Indian Reservation near Chama is host to a number of quality lakes that are prime fisheries at ice out through mid-June and again by late fall. Because it gets pretty hot here during the summer months, the bite will be slow in the high desert lakes but remains fairly consistent in the higher elevation waters.

New Mexico is not without some excellent streams and rivers, many of which are very well known to fly fishermen. The San Juan needs no introduction and is perhaps the best fishery in the state. Add to that the Pecos, Cimarron and the Rio Grande rivers and you have some excellent waters to be explored.

If you concentrate your efforts more in the spring and fall when water temperatures are more favorable to the trout's liking and avoid the hotter summer months when many trout are deep and out of reach of anglers, I'm sure you will not only do better but will enjoy it more.

EAGLE NEST LAKE
New Mexico

New Mexico's Eagles Nest Lake may be insect poor, but thanks to an abundance of Crayfish, rainbows from 2-5 pounds are average as evidenced by this nice 'bow.

Situated below the Sangre De Cristos mountain peaks in northern New Mexico, Eagle Nest lies in an open valley drained by the famous Cimarron River. Trout rich and insect poor, the lake has good numbers of 2 to 5 pound rainbows with fish over 10 pounds possible. Cutthroats from 3 to 6 pounds are also part of the fishery although not as common. The lake doesn't support much in the way of aquatic insects, but the trout do grow large on a thriving population of crayfish.

The lake's abundant crayfish explains why a myriad of large fly patterns that are mistaken for these crustaceans work so well here. Flies such as the peacock and black-bodied Carey Specials in sizes 6 to 10, black Woolly Buggers along with dark brown and black Leeches in sizes 6 to 10, and dark Matukas, Zonkers and Muddlers all in sizes 4 to 6 are all effective.

I found the best time to fish Eagle Nest is right after ice out in the spring and through late June, then again in September and October. Trout can be taken during the summer months, but you will have to work for them. Like most lakes, the weather patterns often dictate fishing success and Eagle Nest is no exception. Because strong winds are common, 9 to 9½ foot, for 6 or 7 weight rods are a good choice to deal with the elements. Most sinking lines will work with emphasis on the intermediate when fishing from shore, or when trout are feeding in the top 4 or 5 feet. Leaders of 9 to 12 feet with tippets to 3x or 4x are recommended to handle the bigger flies and longer casts.

Eagle Nest is not a deep lake and nutrients such as algae can build up during the summer months. With about 2,000 acres of water to cover, boats offer mobility and a little more safety from the shifting winds. Tubing can be effective but care should be exercised when fishing this lake.

Camping is available along the Cimarron River, just below the lake with lodging and fishing information in Eagle Nest and the town of Cimarron.

Dave Freel with a nice 5 pound rainbow taken on an olive body, burnt orange hackled Seal Bugger.

Eagle Nest Summary

SEASON: Open year-round.

TROUT: Rainbow and cutthroat with the rainbows averaging 2 to 5 pounds with fish to 10 pounds possible. The cutthroats, although fewer in number, average 3 to 6 pounds.

LAKE SIZE: Eagle Nest holds about 2,000 acres of water, is shallow and nutrient rich, yet is insect poor.

RECOMMENDED FLY LINES: An intermediate or floating line will work best here unless you fish off the bottom where a type II or uniform sink will work best.

RECOMMENDED FLY PATTERNS: Seal Buggers, Woolly Buggers, Leeches and Carey Specials are consistent with a mix of streamers also working well on these trout.

BEST TIME: Eagle Nest will fish best early and late in the season with ice out a special time for big trout.

STONE LAKE
New Mexico

Stone Lakes rainbows are broad shouldered, aggressive fighters like this 6½ pound male.

The Jicarilla Apache Reservation in northern New Mexico is home to some of the state's top fly fishing lakes. Names like Mundo, Horse and La Jara offer anglers outstanding opportunities for big trout. But, the most prolific concentration of trophy trout is found in Stone Lake.

Only a few short years ago Stone was a carp factory. But in 1993, the lake received an overhaul (poisoning) to eradicate the rough fish. Eight to 10 inch rainbow trout were re-stocked the following year and have prospered ever since.

Today, Stone's trout consistently run 20 to 27 inches and a few push 30 inches. Like most lakes, it takes time to learn her moods and the time tables for insect hatches along with where the trout like to frequent. Time spent on the water will cure most of these mysteries and a lot of other unknowns necessary for consistent success.

Photogenic, it is not. Instead, Stone offers fly fishermen the ultimate challenge for broad-shouldered rainbows. Without an inlet to spawn in, these big trout concentrate in shallow water each spring. Right after ice out is considered by many locals the best time to locate, hook and land one or more of the robust rainbows.

With intense populations of high protein food sources, these Jicarilla Reservation trout grow quickly in this high desert sanctuary. Scuds and leeches account for a big portion of the calories these big fish consume each year. For a change of pace, healthy numbers of midges, damselflies, and dragonflies cram the stomachs of these trout. During the winter months, and into spring, these fish will forage for snails when other insects are sparse. This helps explain why these trout are as big as they are: they are constantly on the prowl for food.

Winter might be considered the lull feeding period. But, in the spring, the hunt intensifies. Following ice out on a mid-May trip here, I fished my black Seal Bugger with purple hackle and my Stillwater Nymph, with equal success. Because of the opportunistic feeding nature of these trout, lots of patterns will work. Local favorites include damsels, dragonfly imitations along with Woolly Buggers and Bunny Leeches both in black, brown and olive.

Fly lines should include floating, Uniform Sink #1 and an intermediate. Leaders can be 10 to 12 feet long with tippets ranging from 3x to 5x. Water conditions and size of pattern should determine the size of tippet. Match your retrieve to the pattern you are fishing and water temperatures. If it is below 50 degrees, go slow. These fish are not going to chase anything when their bodies are still sluggish. When temperatures rise above 50 degrees you can experiment with other speeds, but if you are ever in doubt, go slow.

Most anglers prefer to fish Stone from a float tube or floating device when the trout are concentrated in shallow water. This may be a better time to wade, especially in the spring.

This female 'bow weighed 5½ pounds and was landed with Denny's stillwater system: a Stillwater Nymph, a hand twist retrieve with an intermediate line in four feet of water.

Stone is a big lake covering between 500 and 550 acres depending on water years. The season runs from April 1 to November 30. Opening day can be delayed when ice out still covers the lake. All of the other lakes on the reservation are open year-round. Artificial lures or flies are the only legal methods with a two fish limit in place. Barbed hooks are OK, but bait is not. Licenses (reservation permit) can be obtained at the Jicarilla Apache Tribe Game and Fish Department in Dulce, or at Reel Life in Albuquerque, Suite 10, 1100 San Mateo Blvd. NE (505) 268-1693.

To reach the lake, take I-25 from Albuquerque to Sante Fe, then U.S. 84 through Espanola to Chama, then U.S. 64 west to Dulce. There are no accommodations on the lake.

Stone Lake Summary

SEASON: April 1 through November 30 with ice out prime time for hooking these big trout.

TROUT: Rainbow with the average trout running 20 to 27 inches with fish over 10 pounds possible.

LAKE SIZE: When full, Stone can stretch 550 surface acres of water with plenty of shallow shoreline areas to explore.

RECOMMENDED FLY LINES: I found an intermediate line perfect for shallow water fishing here. When these trout feed shallow, most lines will work. Just match the depth with the lines sink rate.

RECOMMENDED FLY PATTERNS: Local favorites include Woolly Buggers and Bunny Leaches with damselfly and dragonfly nymphs popular as well. I found my Seal Bugger in olive and black good early and late in the day along with my Stillwater Nymph, an excellent imitation for damsel nymph activity mid-day.

BEST TIME: After ice out is magical, but early and late season fishing is fairly consistent.

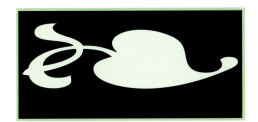

Oregon—This big rainbow was hooked by the author on a small alkaline lake in Eastern Oregon. (Dave Nolte photo)

Chapter 8

Oregon

No other state in the United States, not Colorado, Idaho, Utah, Wyoming, not even Montana can boast of having the numbers of trophy lakes as the state of Oregon. Being my home state, I've obviously had more time and opportunity to explore the stillwater possibilities here compared to those lakes more distant. Make no mistake, this state is loaded. There are at least a dozen lakes that harbor cannibalistic browns or rainbows over 20 pounds.

The mecca for most of these trophy waters lies on the eastern slope of the Cascade Mountain Range centered around the booming town of Bend. Within an hour's drive of Bend you can explore East, Paulina, Wickiup, Crane Prairie, Hosmer and Davis lakes, all of which host trophy rainbows, browns or brook trout. There are several other quality lakes within this web that didn't make my list, but are good fisheries for fly fishermen to enjoy.

Diamond Lake is currently waiting to be poisoned which will eradicate the tui-chub problem. It will then be re-stocked in 2001 taking the first steps to reestablishing it as one of Oregon's best trophy fly fishing lakes. When that happens, Diamond will be deserving of a mention in future lists.

If you travel the eastern half of Oregon you will encounter numerous high desert alkaline lakes such as Mann, Chickahominy and Beulah that are rich in insects and hold good numbers of big trout. These fish not only grow big, but do it quickly with lots of 2 to 5 pound trout available. Almost every little town has its hidden hot spots where local residents fish. Nose around and see what you can find. It might surprise you what lurks beneath the surface in some of these hidden gems.

CRANE PRAIRIE RESERVOIR
Oregon

Crane Prairie Reservoir is one of the most popular lakes in Oregon, and for good reason: it supports lots of big rainbows.

Shallow, intensely rich and infested with big rainbows, it's no accident Crane Prairie is considered one of Oregon's premier trout lakes. With an abundance of aquatic vegetation and old decaying timber stands for cover, the lake once supported dense populations of Callibaetis mayflies, damselflies and dragonflies capable of subsidizing large numbers of bragging size trout. No more! The damsel and Callibaetis populations have all but disappeared because of the illegal introduction of the stickleback minnow that feeds on the larvae of these insects. That's the down side. The flip side is the rainbows prey on this minnow providing anglers an opportunity for some aggressive action using stickleback imitations.

Crane Prairie was formed in 1928 when the Army Corps of Engineers dammed the upper reaches of the Deschutes River. Three main tributaries now feed the lake — Cultus Creek, Quinn River and the Deschutes River. Along with Rock Creek, each of these old river bed channels offer anglers excellent opportunities for large fish during the summer when water temperatures in the main lake heat up. The deepest part of the lake is only 15 feet but most of the trout are landed in depths of 6 to 8 feet.

Rainbows averaging 3 to 5 pounds are numerous throughout the lake but it's the 8 to 12 pounders that test the skill and tackle limits of stillwater fly fishermen. Stories are told and rumors circulate of larger trout landed, and since rainbows to 18 pounds have been taken here, expectations of hooking a big fish are entirely realistic. Brook trout help spice the action and are normally found in the deeper, cooler channels during the summer months. Fish to 6 pounds are possible and occasionally a few large browns are taken.

Anglers can stalk these big trout from the last Saturday in April through the end of October. July through early September are the peak months. Success rates on Crane Prairie are relative to your ability to locate these big trout. In the summer, water temperatures rise concentrating fish in the cooler channels. A host of fly patterns matching the stickleback minnow work on Crane Prairie. Because of the dead timber both above and below the surface, this reservoir used to boast a well-deserved reputation as one of the best damselfly nymph lakes in the country. Now, the focus is not only on the stickleback patterns, but dragonfly nymphs as well. I don't see many anglers fishing with dragonfly nymphs, even though it has become a major food source when the trout can find them. Perhaps this is due to a lack of visual contact with the nymph, but a Carey Special, a Black Diamond or my Seal Bugger in black or olive are excellent matches for this insect. Depending on the time of day and conditions on the lake, a Woolly Bugger or Leech in size 4 to 8, both have their time and place especially when there's a good chop on the water.

A soft tipped, 9 or 9½-foot graphite rod is an excellent choice for these big trout. Fast action rods will not get you past the strike if you use light tippets. Depending on the conditions and pattern being fished, I prefer the intermediate fly line for most situations but a floating or the transparent line is effective as well. Leaders should be 12 to 15 feet with 3x to 4x tippet, but again, patterns and conditions will dictate which to use.

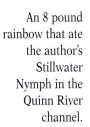

An 8 pound rainbow that ate the author's Stillwater Nymph in the Quinn River channel.

Because of its size, boats are a better way to fish the lake although tubing is popular, but not nearly as effective due to lack of mobility. Wind can be a problem and if it gets to where it's unfishable, there are several other quality lakes nearby, two of which are included in this book, Wickiup and Davis, respectively.

Crane Prairie Reservoir is located about 25 miles west of Highway 97 near LaPine and has several campgrounds around the lake. However, when the bite is on, space can be a problem. Tackle, boat rentals, gas and recreational vehicle hookups are available at Crane Prairie Resort located at the east end of the lake. For fishing information, contact Sunriver Outfitters at (541) 593-8814, or Deschutes River Outfitters (541) 388-8191.

Crane Prairie Summary

SEASON: Opens the last Saturday in April and closes October 31.

TROUT: Rainbow and brook trout with an occasional brown. The rainbows will average 3 to 5 pounds with fish to 10 pounds possible. The brook trout run 14 to 16 inches, but brookies to 6 pounds are prowling the depths daily.

LAKE SIZE: Crane Prairie is a shallow lake with the deepest part only 15 feet. The key is to locate the channels or the springs around some kind of structure.

RECOMMENDED FLY LINES: A floating or intermediate line probably gets the most attention from experienced anglers. When conditions get tough, use a clear line that sinks like a type II.

RECOMMENDED FLY PATTERNS: Minnow imitations that match the stickleback minnow are effective. You will also do well with an olive or black leech, a Woolly Bugger or my Seal Bugger.

BEST TIME: This lake has always been at its best during the warmer summer months. Spring fishing is only so-so, but fall has its moments.

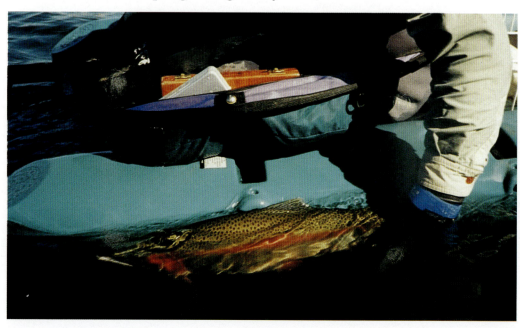

Big trout thrive on Crane Prairie's multiple food sources as evidenced by this 11 pound rainbow being released.

Davis Lake
at dusk is
prime time
for stalking
the big
rainbows.

DAVIS LAKE
Oregon

Formed by volcanic activity over 3,000 years ago when lava blocked the lower reaches of O'Dell Creek, Davis Lake is just one of the many trophy fisheries located off Century Drive in central Oregon. Although not as productive as it was during the 60s and 70s, Davis still offers quality fly fishing for rainbows from 2 to 6 pounds with some monsters over 10 pounds available.

Davis is a shallow lake, but when full it covers 3,900 surface acres. At the deepest spot, near the lava beds on the northeast end, it's 18 feet deep, but the average depth is only about 9 feet.

The lake all but died during the drought in the mid-80s when water levels left it a mere puddle and temperatures rose to dangerous levels. Trout suffered from high pH levels but a fair number managed to survive the low water years. Then heavy snow packs in 1995 and 1996 brought the lake back to normal and along with it came excellent fishing that included some really big rainbows in 1997 and 1998. Fly fishing for these large trout is once again part of the promise of what trophy hunting at Davis is all about.

The lake was designated as a fly fishing only lake in 1939 and is still managed that way today. Shallow and fertile, this is a lake with lots of weed beds and prolific insect hatches that are constant throughout the summer months and well into the fall. A small forage fish called a roach helps sustain these big rainbows along with intense hatches of Callibaetis mayflies, damselflies and dragonflies which are supplemented with midges, leeches and some terrestrials.

Although anglers will use a wide variety of patterns, the most consistent are imitations of Callibaetis nymphs in the spring, damsel nymphs during the summer and dragonfly nymphs fished off the bottom all year.

Black, brown and olive Leeches, along with my Seal Bugger have been deadly for me not only early and late in the season, but work well early and late in the day throughout the season. During the summer months, I find my Stillwater Nymph, peacock Callibaetis Nymph and my A.P. Emerger very consistent on Davis rainbows. But, don't hesitate to try your favorite patterns. Most flies presented properly and fished at the right depth work well.

Although Davis Lake is home to lots of large trout, double digit rainbows like this 10 pounder are a minority.

Because this is a shallow lake, I use the intermediate line, yet I've witnessed just about every line available in use here. The neat thing about Davis Lake is you can play the game with lots of combinations of lines, flies and retrieves and most will work when these trout are feeding. Those anglers who out-fish the rest of the clan obviously know the lake, but more importantly concentrate their efforts where these fish feed. The most productive areas are off the campgrounds where O'Dell Creek enters and by the lava rock area at the far end of the lake.

A nutrient rich lake, Davis gets an algae bloom during the summer months forcing the trout to take refuge in the cooler temperatures of the O'Dell Creek channel. But, good numbers of big trout are found holding off the deeper water by the lava rock area as well. You will find these rainbows scattered throughout the lake, but more concentrated in those two areas.

The season on Davis never closes so you can work this body of water all year or until it freezes up in late November. Davis will usually shake its wintry blanket by late March, or sooner if it warms or the wind blows.

Davis has a one fish limit and prohibits fishing from a motorized craft although it is legal to move from one spot to another with a motor.

I've never found a particularly bad time to fish this lake, but the most consistent periods for me have been early mornings and early afternoons from June through September. Some of the bigger fish are taken right after ice out, but be prepared for cold temperatures that occur in early spring.

There are three campgrounds located on the lake, each with a boat ramp. For accommodations and supplies, there are several resorts in the area. For fishing information and updates, contact Sunriver Outfitters at (541) 593-8814, or Deschutes River Outfitters (541) 388-8191.

This 10½ pound female rainbow is the largest fish the author has landed from Davis and was taken on an olive Seal Bugger in 1997. (Dave Nolte photo)

Davis Lake Summary

SEASON: Open year-round.

TROUT: Rainbows with most fish running 2 to 6 pounds and some reaching 10 or more pounds.

LAKE SIZE: Davis is a shallow lake covering 3,900 surface acres.

RECOMMENDED FLY LINES: I prefer the intermediate for most of my fishing and will use a clear full sinking line that sinks about 2 inches per second early in the season when trout tend to hug the warmer, yet deeper water.

RECOMMENDED FLY PATTERNS: Whatever pattern you choose, a black or olive one will get you results on Davis. My Seal Bugger is very good early or late in the season or during those early or late periods when trout like to feed. My Stillwater Nymph, along with my peacock or grey Callibaetis Nymph does very well when hatches are occurring and trout are feeding on or near the surface.

BEST TIMES: Davis is good all season, but you have a better chance of hooking the bigger trout here in the spring and again in late fall.

EAST LAKE
Oregon

East Lake
is one of
Oregon's
best trophy
fly fishing lakes
where browns
like this 6
pounder are
fairly
common.
(Dave Nolte
photo)

A natural lake situated within the cone of Newberry Crater, East Lake is another of the many trophy lakes found in the state of Oregon. It's located about 35 miles south of Bend at an altitude of about 6,300 feet. This is a deep lake, yet offers excellent shoal fishing in the spring and fall when the water temperatures have cooled. During the summer months the bigger trout go deep, yet fly fishing remains fairly constant for smaller fish near the shallows and along the edges of the weed beds.

Rainbows and browns to 10 pounds are possible with a few monster browns to 20 pounds still present in the lake. It is thought by many that East Lake is one of three lakes in central Oregon that possibly holds the next state record brown trout. There are plenty of rainbows and browns from 4 to 6 pounds with the browns more available during the fall months. East was once an outstanding brook trout fishery with lots of fish to 4 pounds present, but their numbers have dwindled in past years. East Lake also boasts a kokanee and Atlantic salmon population. First introduced in 1990, Atlantic salmon here run between 1 and 2 pounds. These are long and lean fish, scrappy fighters and will attack a well-presented fly.

A Callibaetis hatch usually comes off beginning in late spring extending into fall. During this period nymphing with most standard patterns in sizes 10

There are plenty of big rainbows in East Lake waiting to be challenged.

to 14 works well, however a good presentation is a must. As with most lakes that harbor big fish, Leeches, Woolly Buggers and a variety of multi-colored minnow imitations will all take fish. I've had some exceptional action using my A.P. Emerger, Callibaetis and Stillwater nymphs along the weed bed edges and drop-offs leading to deeper water. I've used my Seal Bugger in olive with a burnt-orange hackle to search out the big browns and enjoyed my best results early and late in the season just before dark. During the spring, particularly after ice out, rainbows and browns can be found prowling the shallow water from 4 to 6 feet deep.

The same pattern occurs in the fall when the browns once again move from deep water to the shallow water. Anglers fishing with fast sinking lines who jig their weighted fly patterns up off the bottom where the gradients slope can take some big fish. Finding feeding fish is often more of a concern than getting them to take.

You will need an intermediate line for working the shallows or when fish are feeding just below the surface down to about 4

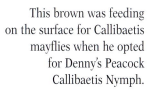

This brown was feeding on the surface for Callibaetis mayflies when he opted for Denny's Peacock Callibaetis Nymph.

feet. Type II and III lines are better for probing deeper water from 8 to 15 feet. The uniform sink lines are ideal for this type of fishing. Patterns should be weighted at the head to get the undulating motion that entices these big trout to hit.

The lake opens to fishing on the last Saturday in April and closes the end of October. Because of altitude (6,370 feet), East can be highly subject to cold and stormy weather patterns passing through. Freezing temperatures are common in the spring and fall, but the anticipation of hooking one of these big trout will make you forget about the discomforts.

A boat offers the best mobility and you can use it to reach certain areas and then fish it from shore. Tubing is also popular, but wading is seldom effective at East.

There are campgrounds on the lake along with boat rentals. Fishing information is available at East Lake Resort located on the lake, or you can call Sunriver Outfitters (541) 593-8814.

East Lake Summary

SEASON: East Lake opens the last Saturday in April and closes October 31.

TROUT: Rainbow and brown trout, Atlantic salmon with an occasional brook trout. The rainbows run 2 to 4 pounds with plenty of 4 to 6 pound fish lurking in the depths. The browns average 3 to 6 pounds with some lunkers over 10 pounds available and fish to 20 pounds possible.

LAKE SIZE: This is a deep lake up to 160 feet, but the best fishing is along the weed bed edges or near the drop-offs. There is plenty of room to test your skills with 1,044 acres.

RECOMMENDED FLY LINES: When fishing shallow or when trout are working on top, I prefer the intermediate line and switch to a clear type II full sinking line when trout are deeper.

RECOMMENDED FLY PATTERNS: Lots of stuff works here. I like my Seal Bugger to hunt with, especially during the late hours when browns cruise shallow water. When hatches occur, I prefer my A.P. Emerger, Callibaetis Nymph or Stillwater Nymph.

BEST TIMES: The lake fishes well all year, but is best for the bigger trout early and late in the season. There are good numbers of big trout here that will help you keep your focus.

Hosmer Lake, with its tranquil setting, supports the only Atlantic Salmon fishery in the West and is also home to some very large brook trout. (Dave Nolte photo)

HOSMER LAKE
Oregon

Located in the Deschutes National Forest, Hosmer Lake is another of the quality lakes nestled among the Cascade Mountains in central Oregon. But, unlike the others, Hosmer is one of a handful of lakes in the western United States that contain landlocked Atlantic salmon. At Hosmer they share the same water with some of the largest brook trout found anywhere in the state, not to mention the hordes of anglers who pursue them.

Located on Century Drive about 35 miles southwest of Bend, Hosmer is restricted to fly fishing only with barbless hooks. The lake covers about 150 acres divided into an upper and lower lake. A small channel separates the two. The average depth is about 10 feet making it perfect fly fishing water. Wading is impossible, but fishing from float tubes, canoes, small prams or boats is ideal. Only electric motors are permitted which is the way it should be in this tranquil mountain setting. Regulations require anglers to release all Atlantic salmon unharmed.

Open year-round, Hosmer's waters are clear and cold allowing anglers with relatively good eye sight to spot fish cruising or just being lazy near the bottom. This is especially true in the channel where you will see astonishingly large brook trout hugging the lake bottom.

Dave Nolte releases a nice 4 pound Atlantic Salmon from Hosmer's pristine water.

Years ago the Atlantic salmon ran 5 to 7 pounds, but a 15 to 20 inch fish is closer to the average today. The brook trout, however, are another story. These trout run 2 to 6 pounds and are anything but easy. I still consider a big male brookie the toughest of all trout to catch. Perhaps it is their shy nature, or the fact that they prefer structure near the bottom that keeps them out of reach of so many anglers.

The lake is lined with tules, cattails, willows and aquatic vegetation all of which form the perfect habitat for these fish. It also makes for good hatches of Callibaetis mayflies beginning in early June immediately followed by damselflies which last from mid-June through August. Add to that healthy numbers of leeches and dragonfly nymphs for these salmon and trout to munch on throughout the season and you'll know why these big fish are large and feisty.

To match up, I've had my best action using my Callibaetis Nymph or A.P. Emerger in grey or olive in size 12 when the mayflies are present and my Stillwater Nymph in size 12 when the damsels begin to show. If there is no surface activity taking place, I'll go to a size 10 or 12 black Leech or my black Dragon Nymph fished off the bottom. Hares Ear or Pheasant Tail nymphs are also good choices when trout or salmon are showing.

A nice 16 inch male brookie taken on Denny's Olive Callibaetis Nymph in early October. (Dave Nolte photo)

Most of the time I'll use an intermediate line, but if I need to dredge near the bottom for the big brookies, a fast sinking line is a better choice. When there is good topwater activity, use a floater to cover the surface feeders. To be successful on the bigger brookies and Atlantics, you will need to make errorless presentations. Use long leaders, 12 to 15 feet, and light tippets, 5x or 6x. That is especially important if the water is flat. When conditions are difficult, such as a high sun and flat water, try using the clear transparent line with a very slow retrieve.

The best time to challenge these fish is early June through mid-July and again in mid- to late-September. August is usually on the slow side, as the fish tend to lie near the bottom and wait out the warm, hot summer days. In the fall, a popular method is to quickly strip white or yellow streamer patterns. The fast retrieve, combined with these colorful patterns, can trigger savage strikes from the Atlantic salmon.

There are two campgrounds on the southern lake and there is a boat ramp on the southern end. Lodging is available at several nearby resorts in the area. You can get up to date information by contacting the Sunriver Outfitters at (541) 593-8814.

Hosmer Lake Summary

SEASON: The lake is open year-round, however heavy snow pack can keep anglers from gaining access until late April.

TROUT: Atlantic salmon and brook trout. The salmon run 16 to 20 inches with an occasional bigger fish. The brook trout run 2 to 6 pounds and I rarely catch one smaller.

LAKE SIZE: You have 150 acres to fish in with both the upper and lower lakes equally good. This is not a deep lake. The average depth is about 10 feet but a boat or float tube is necessary to reach these fish.

RECOMMENDED FLY LINES: For surface activity use a floating line, an intermediate line when fishing sub-surface or a type II fast sinking line for probing the bottom for brook trout.

RECOMMENDED FLY PATTERNS: I like my Callibaetis Nymph and A.P. Emerger Nymph for fishing the Callibaetis mayfly hatch. My Stillwater Nymph is effective for the damsel hatch and my black Leech or Dragon Nymph when working the bottom structure for the brookies. Most standard nymphs will work just fine here. The key is the depth and angle you fish them.

BEST TIMES: This is an early spring or late fall lake with mid summer only so-so. Key on the hatches whenever fish are feeding on top.

Upper Klamath Lake may be the best trophy trout lake for fly fishermen in the country.

UPPER KLAMATH LAKE/AGENCY LAKE
Oregon

Shallow, nutrient rich and extremely productive in terms of producing big trout, Upper Klamath Lake is considered by many to be the best trophy rainbow trout lake for fly fishermen in the country. I'd have to agree.

Located in southern Oregon, Upper Klamath Lake is huge. It sprawls over 64,000 acres and is 26 miles long. Agency Lake, with 8,200 acres by itself, is connected to Upper Klamath at the north end of the lake by a half-mile long channel giving anglers over 100 miles of shoreline to cover while pursuing the huge fish that inhabit these two lakes.

Trout grow big and fast in this food rich environment with the average fish running 18 to 23 inches and a whopping 3 to 6 pounds. Although the majority of the trout landed are 3 to 8 pounds, it's the 10 and 15 pound trophies that keep anglers coming back. In years past, monster rainbows from 18 to 25 pounds were not at all uncommon. Still, fish between 18 and 20 pounds are hooked occasionally, but few are landed for anglers to brag about.

Agency and Upper Klamath lakes are fed by seven tributaries, the Wood and Williamson rivers on the eastern shore, and Crystal, Recreation, Harriman, Short and Odessa creeks on the western shore. Of these the Wood and Williamson rivers have well-deserved reputations as trophy fisheries by themselves. Each summer these rivers are host to large schools of migrating rainbows seeking the cooler water temperatures. The same is true for the springs within Upper Klamath Lake's Pelican Bay.

This 14 pound rainbow the author landed was hooked in five feet of water with a black Seal Bugger off the mouth of Crystal Creek.

The underwater habitat in Pelican Bay is a favorite of migrating rainbows seeking cooler temperatures from the main lake. (Dave Nolte photo)

Upper Klamath and Agency Lake are entirely dependent on natural reproduction to replenish their numbers and are two of only a few lakes in the United States to support an entire fishery of wild, native trout. Attempts to introduce hatchery trout have failed in the past due to a disease organism present within the lake. Two genetically different species of rainbows along with a rainbow/steelhead cross comprise the lake's trout family. Some large browns and a few brook trout are present in some of the tributaries, especially the Wood and Williamson rivers, but are rarely taken in the main lake.

Upper Klamath's rainbows add weight quickly like this three year old female that measured 23 inches and weighed about 6 pounds.

With an average depth of only 7 feet, there is an abundance of food sources for trout to pick and choose from. During spring and early summer, small forage fish and

Agency Lake off the mouth of the Wood River is a favorite of anglers and trout alike.

leeches comprise the bulk of what these big trout eat. As summer water temperatures heat up, midges, mayflies, caddisflies, damselflies and dragonfly nymphs, as well as scuds, crayfish and a few terrestrials, all help supplement their diet.

My Seal Bugger originated on Upper Klamath and even today I still consider it my best pattern for these big fish. I prefer the black, brown and olive Seal Buggers, along with Leeches in sizes 6 or 8. My Stillwater, Callibaetis and Black Diamond nymphs is sizes 10 to 14, plus the standard nymphs such as the Zug Bug, Pheasant Tail, Prince and Hares Ear nymphs are excellent for fishing the flats, inlet mouths

Big trout like this 8 pound rainbow are common in Agency Lake.

This rainbow estimated at 7-8 pounds never saw the net, but twenty minutes later a bigger one did.

and weedy areas. Assorted streamer patterns in white, olive and brown are also effective during the spring and fall months, but I seldom see other anglers using them except when trolling. My Shiner Minnow and Chub Minnow patterns also originated on Upper Klamath and are deadly when these big fish are looking for forage fish.

You will manage these big trout best with a 9-9½ graphite rod matched for a 6 or 7 weight line. My custom rods with strong butt sections and soft tips were designed to handle the big trout and strong winds that are so much a part of fly fishing Upper Klamath Lake.

Fly fishermen should consider using the intermediate fly line for the majority of their fishing here. When fishing the creek inlets or Pelican Bay, the clear transparent line or the type II uniform sink lines are also good choices for exploring the spring and weed bed areas. It is wise to remember that line selection is relative to patterns and depths you're fishing.

Unlike most lakes, Agency and Upper Klamath peak during the warmest periods of the year with July and August being prime time. During these months, increasing water temperatures in the main lakes force trout to concentrate near cooler water, thus location becomes more of an issue. Prime areas during this time include the mouth of the Wood River where it enters Agency Lake, the Williamson River where it enters Upper Klamath Lake, Pelican Bay and the tributaries near Rocky Point along with the Fish Bank area off the mouth of Pelican Bay.

Because of the lake's size, mobility is critical. Boats with large motors are a must. Float tubers are limited in range and wading is impossible because of access and the lake's marshy habitat. Winds are a factor, especially during spring months and boaters should exercise caution when venturing out on the main lake.

U.S. Forest Service Campgrounds are limited around the lake, but Rocky Point Resort, a full service facility, offers cabins, recreational vehicle hookups and tent sites along with boat rentals, tackle and restaurant.

For some unknown reason, Upper Klamath and Agency lakes remain one of the best kept secrets for trophy trout in the country. Yes, it can be moody at times like most lakes, but the rewards are worth the energy spent once you discover some of its secrets. I know, because I have fished and guided these two lakes the past 25 years.

Upper Klamath/Agency Lakes Summary

SEASON: Open year-round.

TROUT: Rainbow with an occasional brown and brook trout showing near the tributaries. The rainbows average 3 to 6 pounds with double-digit trout very common. Some monsters exceeding 15, up to 20 pounds are still there just waiting for your challenge.

LAKE SIZE: Upper Klamath and Agency lakes combined exceed 100 miles of shoreline, yet the average depth is only about 7 feet. Upper Klamath is 64,000 acres while Agency is 8,000 acres.

RECOMMENDED FLY LINES: I find the best line for fishing the main lake is the intermediate slow sinking line.

RECOMMENDED FLY PATTERNS: For fishing the main lake, I use my Seal Bugger or Leech patterns in black. When trout are working minnows, most streamer flies will work. I switch to my Chub or Shiner Minnows for this type of feeding activity. When fishing in or near the bays or tributaries, I like my Stillwater Nymph or my Callibaetis Nymph when these big fish feed near the surface.

BEST TIMES: These lakes both fish best from mid-June through September with July and August being peak for the biggest trout as they move into cooler water.

Pattern selection
is usually not
critical at
Mann Lake,
but the trout
will tell you
if you made
the right
choice.
(Dave Nolte
photo)

MANN LAKE
Oregon

If there was ever a lake situated in the middle of nowhere that is worth searching out, then Mann Lake is it. Located at the base of the Steens Mountains in the southeast corner of Oregon, Mann Lake is a typical high desert alkaline fishery. It is extremely rich in nutrients, has an abundance of aquatic insects and hefty Lahontan cutthroats to boot. Its reputation isn't so much for the size of its trout, but for the quality fly fishing. Some believe the area is better suited to antelope, jackrabbits and rattlesnakes, but looking up at the grandeur of the Steens, the setting is quite scenic and no one complains about the solitude and clean air of a desert setting.

Managed as a brood stock lake, cutthroats and rainbow-cutthroat hybrids average 2 to 3 pounds with some exceeding 5 pounds. Like most alkaline rich lakes, Mann is shallow averaging 8 feet in depth. It covers about 275 acres and flourishes with aquatic vegetation. Open year-round, Mann supports a barbless hook regulation and a two fish limit, 16 inches or better. If you do choose to keep one, don't expect a gourmet flavor.

From a fly fishing perspective, the lake is infested with a mix of aquatic insects including midges, Callibaetis mayflies, dragonfly and damsel nymphs. However, these trout are rarely selective and most standard nymphs will work.

The author plays a nice cutthroat that was cruising shallow water. (Dave Nolte photo)

This 18 inch cutthroat is typical of what you can expect from Mann Lake.

I prefer my Stillwater and Callibaetis nymphs, but a Hare's Ear, Pheasant Tail, Prince Nymph and Carey Special in sizes 10 to 14 are all consistent during the summer months, especially when mayflies and damsels are on the water. My Seal Bugger, and leech patterns along with Woolly Buggers in sizes 8 to 10 are deadly right after ice out in the spring, then again in September and October when insect hatches begin to diminish. This is the ideal lake to try your fly tying fantasies. Fish those wild creations you marveled at, but scared the heck out of fish in other lakes. They might just work here.

Both floating and intermediate lines will match the conditions you'll encounter, and since these trout aren't real fussy eaters, leaders from 9 to 12 feet tapered down to 4x or 5x tippets work just fine.

I've never found a bad time to fish this lake, although many believe the bite slows during the summer months when the higher water temperatures ignite algae blooms. I actually prefer July, August and early September when the trout are on a diet of aquatic insects, although spring and fall are consistently good if the wind doesn't blow too hard.

First light is an excellent time to sight fish cruising trout working the shallows before the morning sun moves them into deeper water. I've always enjoyed excellent early morning action using a size 14 tan Callibaetis Nymph or a size 12 olive Stillwater Nymph. Use a slow, hand-twist retrieve when wading the edges and sight fishing cruising trout. Be sure to remove your line softly from the water after a refusal and don't line cruising fish. They may dart from the area alerting other fish to your presence. Remember, careful wading and a good presentation are prerequisites to consistent action when stalking any trout feeding in shallow water. As the sun rises, these fish will slowly retreat to deeper water hanging around weedy areas while continuing to feed. It's a great lake to hone your skills and boost your ego.

As I said, expect algae blooms during the summer months. Water quality may suffer a bit, especially if persistent winds keep nutrients mixed. The use of a small boat, float tube or wading are all effective, but a boat allows you to cover more water.

Since there are no accommodations at Mann, you'll need to bring your own food, water and camping unit. A few spare tires aren't a bad idea either since flats are common on the hard shale road leading to the lake.

Mann Lake Summary

SEASON: Open year-round.

TROUT: Lahontan cutthroat and rainbow-cutthroat hybrids with the average trout running 2 to 3 pounds with some exceeding 5 pounds.

LAKE SIZE: Mann Lake has about 275 acres of water, is extremely shallow with lots of quality shoreline habitat to fish.

RECOMMENDED FLY LINES: All you need for this one is an intermediate line. A floating line is needed to fish dry flies, but is not recommended when heavy winds force drag into your presentation.

RECOMMENDED FLY PATTERNS: I do extremely well here with my Stillwater or Callibaetis nymphs during the warmer months, or with my Seal Bugger early or late in the season.

BEST TIMES: Many anglers shy away from Mann Lake during the summer months because of the algae, but I have always found June through September to be very good. There really isn't a bad time here, but you have to expect windy periods anytime you fish here especially during the spring months.

MILLER LAKE
Oregon

Deep, gin clear lakes are not generally very good waters for fly fishermen. I usually avoid them because they don't fish well with flies. Miller Lake is an exception. Despite a narrow window of opportunity, when fishing borders on fabulous, this lake's trophy trout can be caught in shallow water when they move there to feed. This presents a challenge worth exploring.

Miller Lake contains mostly brown trout and kokanee salmon sharing their space with numbers of small rainbows. Most of the fishing pressure is from trollers, some preferring to chase the kokanee, others pursuing the big browns. As a result of netting operations to determine growth rates, studies conducted by the Oregon Department of Fish and Wildlife show healthy numbers of large fish. Roger Smith, a biologist with the ODFW, found a high percentage of 5 to 7 pound fish as well as large numbers of 8 to 12 pound browns. These tests also indicate the rainbows, for whatever reason, are not doing as well in terms of adding weight.

From time to time, trollers will land browns to 15 pounds, and they are quick to tell you tales of bigger fish that were lost. That is one of the reasons I like Miller Lake. You get your share of stories and rumors of big trout lurking in the depths of this lake. Somehow, Miller has escaped the fraternity of

Miller Lake is one of Oregon's unknown trophy lakes that supports good numbers of double digit browns.

anglers carrying fly rods. I have seen only one other fly fisherman challenging these big trout in past years and he didn't appear to be a threat to the fishery.

Miller is surrounded by steep slopes with thick stands of pine trees. It is a pristine body of water with 4 1/2 miles of shoreline and a maximum depth of 145 feet. Because there are no weedy areas to support insect hatches other than sparse numbers of midges, the primary food source for these big trout is kokanee fry.

Whenever I fished Miller Lake, I had my best luck using my olive Chub Minnow and Seal Bugger in sizes 8 or 10. I've also done well with my Stillwater and Callibaetis nymphs in size 10 for browns up to 7 pounds. Remember, the emphasis is on presentation, not so much on pattern. There will be little margin for error when stalking these big trout in shallow water. I found the best response using a slow retrieve or a short jerky pull when I was getting bumps but not hook ups.

Miller is rarely a numbers lake, but you only need one of these big monsters to keep you wired for a few days. I've always found a little wind and an overcast day best.

Even though spring and fall are the most productive times to fish Miller, you have a shot at these fish any day that you can be there before sunrise. Browns are notorious nocturnal feeders and your key to success is to be there when they frequent the lake's shallow shoreline areas. During the fall, these trout often remain along shoreline edges for much longer periods moving into deep water only when threatened.

If you choose to pursue these big trout when they are holding in deep water, you are not going to be very successful. Without weed beds for cover, the browns must rely on downed timber that litters the shorelines around the lake when they hunt for food in shallow water. This, at least, will provide anglers with some clues as to where to stalk these big fish. Other prime areas are near the two little inlets that feed Miller, Tipsoo and Evening creeks. Obviously, trout will be there during spawning time, but will also frequent shallow shoreline areas when looking for spawning gravel.

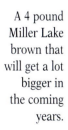

A 4 pound Miller Lake brown that will get a lot bigger in the coming years.

Because this is clear water, the best fly line for Miller is the transparent line followed by the intermediate line. Fast sinking lines are fine for working the ledges where drop-offs occur, but you need to determine the depth they are holding and this zone is often narrow. Patience is certainly a virtue if you fish deep water.

Miller is located about 12 miles off Highway 97 just north of Chemult in south-central Oregon. There is no closed season here, but the road in is not maintained when it snows. Ice out can be the best fishing of the year, if you can get in. During the late spring and summer months, mosquitoes are a problem so go prepared with bug spray. There is camping and a boat launch on the lake. Lodging and supplies are in Chemult.

Miller Lake Summary

SEASON: Open year-round.

TROUT: Browns and rainbows with a high percentage of browns running 5 to 7 pounds and 8 to 12 pound fish occasionally showing on the stringers of trollers.

LAKE SIZE: This is not your classic stillwater lake. Most of the lake is over 100 feet deep, but there are 4 1/2 miles of shoreline littered with downed timber and shoal areas for trout to cruise.

RECOMMENDED FLY LINES: The clear transparent lines will fish this lake best. When trout are in shallow water, you will need long leaders if you choose to use a colored line. If the trout are not showing in the shallows, use a type II sinking line and work the drop-offs.

RECOMMENDED FLY PATTERNS: These fish feed heavily on kokanee fry so think streamer-type flies. I have had my best action with my olive Chub Minnow and an olive or black Seal Bugger. When trout show in shallow water, try my Stillwater or Callibaetis nymphs.

BEST TIMES: From a seasonal standpoint, I prefer early spring and late fall. Accessing the lake due to unplowed roads can be a problem. During the summer months, very early or late in the day can bring these big browns into shallow water.

Paulina Lake
is a popular
lake for trollers,
but is seldom
challenged by
stillwater fly
fishermen.
(Dave Nolte
photo)

PAULINA LAKE
Oregon

Situated in an area where trophy trout lakes flourish, Paulina Lake goes relatively unnoticed by fly fishermen. Part of Paulina's disguise is its size. It sprawls a breathtaking 1,500 acres, and because of its location, in the middle of Newberry Crater, it is also a deep lake with limited shallows, weed beds or aquatic insects.

Most of the fishermen who challenge the brown trout in Paulina troll early in the year immediately after ice out when the big browns are prowling the shoreline. In 1993 Guy Carl caught the state record brown trout surface jigging a Rapala in shallow water. That fish weighed 27-pounds, 6 ounces. In 1965, a world record brown was taken from Paulina weighing 35 pounds, 8 ounces, but it wasn't landed with a rod and reel nullifying any possible record. It does illustrate the point, however, that this lake grows incredibly large brown trout. Surprisingly, with fish this size, the lake doesn't attract a lot of attention from fly fishermen.

For me, fishing this lake over the years was an adventure, very much an unproductive period for the most part. But in time, lessons were learned and much of the mystery was removed. If there is a negative here it is that the lake presents a limited window of opportunity. Because it is a deep lake, it's a waste of time to challenge these big fish from late June through mid-September when they are hugging the depths. Trollers might reach them, but not fly fishermen — not unless you're lucky enough to find one cruising the shallow water early or late in the day. But from ice out to early June, and again in the fall, these monsters become vulnerable when they enter the shallow weedy areas to feed. Water temperatures are more favorable then, not to mention their biological need to reproduce each fall. Without a natural inlet to spawn in, many will seek the shallow areas although space is limited.

The double-digit browns are fewer now, although you can expect good numbers of them from 4 to 8 pounds. There are also a lot of rainbows in Paulina, but fish over 5 pounds are rare. Kokanee too grow large in this lake; 2 and 3 pound fish are common although difficult to take on flies.

This 6 pound brown was taken in less than three feet of water in late October.

Pattern selection is seldom a priority, but presentation is. Because these big fish rely on kokanee fry to sustain their weight, streamers and Seal Buggers are my first choices, but I've enjoyed some excellent action using my Stillwater and Callibaetis nymphs as well. A big key for success here is the depth and speed of retrieve you choose to fish your flies. Because I prefer to challenge these big fish when they enter shallow water, I like the clear transparent or the intermediate lines. I'll concentrate my efforts in the top 12 inches or just off the bottom in water 6 feet or less. Move your flies slowly using a long, 2 foot retrieve with minnow imitations or a Seal Bugger. If you use smaller patterns, try a hand-twist retrieve.

The best time, regardless of time of year, is just before the sun rises and again just before dark. These big trout are shy creatures, cautious and on full alert when they prowl the shallow water. That means your presentation must be flawless.

At an elevation of 6,331 feet, winter seems to hang on longer here, lasting into the early spring. But when ice out occurs, it's important to be there. It can be the best fishing of the year.

There are three campgrounds on the lake along with Paulina Lake Resort. Lodging is available in Bend about 35 miles away.

To get to the lake from Bend go south on Highway 97 for about 20 miles. Turn onto County Road 21 and drive 13 miles. Bring warm clothes.

Paulina Lake Summary

SEASON: Paulina opens the last Saturday in April and runs through October 31.

TROUT: Browns and rainbows with the average brown running 3 to 5 pounds with 4 to 8 pound fish common and some monsters exceeding 20 pounds lurking in the depths. The rainbows top out at 5 to 6 pounds with most running 12 to 15 inches.

LAKE SIZE: This one is big and deep with over 1,500 acres and very little shoal area for stillwater anglers to explore. Nevertheless, these big browns will come in to feed in the spring and spawn in the fall.

RECOMMENDED FLY LINES: An intermediate or transparent line will fish best in the shallow areas. Fast sinking lines are a good choice when these big fish hold near drop-offs close to deep water.

RECOMMENDED FLY PATTERNS: Streamers and Seal Buggers work best when exploring the shallows just before first light and again before dark. When fish are working near the surface, I use my Stillwater and Callibaetis nymphs.

BEST TIMES: From ice out to mid-June and again late fall are prime times on Paulina. Early morning and again in late evening can be good when fish cruise the shallows.

Wickiup Reservoir at dawn.

WICKIUP RESERVOIR
Oregon

Most stillwater fly fishermen don't know that Wickiup Reservoir in central Oregon is a premier trophy lake. For years, trollers and bait fishermen have been getting all the headlines. A large number of anglers landed big browns from Wickiup and have been eager to talk about it. And why not? They have their fish to prove it. Oregon's state record brown trout was challenged on opening day in 1998, but missed the mark by a few ounces. Had the angler weighed it immediately it probably would've surpassed Guy Carl's record still held at Paulina Lake. But it stayed in a cooler for the better part of the day and undoubtedly lost some weight.

Despite these kinds of stories, Wickiup remains a mystery for most fly fishermen. Perhaps the reputations of lakes like Davis, Crane Prairie, Hosmer, East and Paulina, all of which are within short driving distance from Wickiup, keeps the fly fishing pressure off this lake. Or, could it be the reputation of the brown trout being a difficult adversary to catch intimidates stillwater anglers. Maybe it's the size of Wickiup that helps disguise her secrets that overwhelms fly fishermen. Whatever the reason, anglers toting fly rods are a minority here.

Being a reservoir, Wickiup can be moody at times. Summer draw down keeps water levels constantly changing, forcing these trout to search for new cover. Although plentiful at times, insect hatches are inconsistent. Only in heavy snow pack years when water levels fluctuate little are aquatic insects found in decent numbers to spur sub-surface feeding.

Mayflies, caddisflies, midges and damselflies help sustain the trout in Wickiup, but the main source of protein for these trout comes from leeches, crayfish and an abundance of forage fish, in particular, kokanee fry.

Although the main quarry here is the brown trout, there are almost equal numbers of rainbow trout. Add to that a growing population of brook trout, kokanee and whitefish, and anglers who fish this lake can be kept busy. Browns to 20 pounds are not that uncommon, but the average fish runs 2 to 6 pounds with 8 to 10 pound fish showing on occasion. The rainbows are some of the hardest fighting trout I have ever encountered with 12 to 15 inches more the average; however, 'bows to 8 pounds are possible. In the last few years the brook trout have been growing in numbers and size. Brookies to 17 inches averaging 2 to 3 pounds were landed in 1998 and should continue to add weight in the years ahead.

Much of the difficulty in catching big trout in Wickiup comes from anglers' inability to locate them. With an abundance of inlets and bays, deep water and surface acres to explore, Wickiup needs to be scouted, with time spent fishing to get to know her moods. It's not a particularly deep lake; most of its shoreline areas are shallow. But it is a lake that needs to be fished from a boat or float tube if you want to try to find these fish.

As a rule, each spring and fall the bigger browns prowl the shallow water for food or to prepare for spawning. There are two main tributaries into Wickiup, the Deschutes River and Browns Creek. Both close August 31 to pro-

Dave Nolte with a 4½ pound brown from Wickiup Reservoir.

tect migrating browns following the kokanee that move up into these tributaries to spawn.

For me, pattern selection has never been critical and lots of flies are acceptable to these trout. A variety of minnow patterns will work on these big browns. I prefer my olive or black Seal Bugger, along with my Chub and Shiner Minnow patterns. All have done plenty of damage in past years.

Browns and brook trout are more inclined to feed nocturnally, although midday can be very good when insects and a riffle appear on the water. Its during these mid-day feeding binges that I prefer my Stillwater or Callibaetis nymphs or my A.P. Emerger Nymph in size 10. But again, lots of flies work so concentrate on depth and retrieve and experiment until you find something that does the job. I almost always use the intermediate or clear transparent line on Wickiup and fish the shallow areas close to shore to intercept cruising browns early or late in the day. I don't see a need for fast sinking lines for this style of fishing, but if you want to probe the depths, then try it.

Leaders and tippets are very much relative to pattern selection and conditions. If it's flat, fish with longer, lighter tippets and smaller patterns. Early or late in the day when the there is surface chop, I can get by with 10 to 12 foot leaders with 3x or 4x tippets. If the water is clear I'll use the fluorocarbon tippet material which seems to draw more strikes, all other factors being equal.

There is an abundance of campgrounds sprinkled throughout the area and a few lodges available for other supplies. Wickiup is located off Century Drive which leads to nearby North Twin and South Twin lakes. Wickiup is a 30 minute drive from LaPine off Highway 97, and about 45 minutes from Bend. With so many other good lakes in the immediate area, if Wickiup is slow, try one of the others. Chances are good one of them will be producing well.

Brook trout, once a real minority in Wickiup, are now showing up in good numbers with some reaching 3 pounds.

Wickiup Reservoir Summary

SEASON: Opens the last Saturday in April and closes the end of October.

TROUT: You will find a mixed bag of brown, rainbow, brook and kokanee. The brown trout can be huge with fish to 26 pounds landed in 1998. The average brown runs 2 to 6 pounds with good supply of 8 to 12 pounders available. The rainbows are in good numbers with some fish to 8 pounds possible. Brookies are gaining in numbers each year with fish to 4 pounds. The average brook trout runs 12 to 14 inches.

LAKE SIZE: Wickiup is big and spread out with bays and fingers stretching out in all directions. Summer draw down keeps the water levels constantly changing and the size never the same. But this is a quality big fish lake with plenty of old streambed channels to explore.

RECOMMENDED FLY LINES: You can do well with fast sinkers working just off the bottom or an intermediate line to fish shallow water. The clear transparent line is a good choice when the surface water is flat and no activity showing. A floating line comes in handy when hatches bring trout to the surface.

RECOMMENDED FLY PATTERNS: My Seal Bugger or small streamers work well early or late in the day, or season. When I work over feeding trout near the surface, I prefer my Stillwater, Callibaetis or A.P. Emerger nymphs in size 10. Many quality nymph patterns will work here if you keep them in the feeding zone.

BEST TIMES: This lake fishes well all season, but is best for the bigger trout in spring and late fall. Early morning and late evenings will be best for the big browns and rainbows.

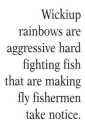

Wickiup rainbows are aggressive hard fighting fish that are making fly fishermen take notice.

Dave Freel with a 7 pound rainbow from Utah's Strawberry Reservoir.

Chapter 9

Utah

Utah is a state relatively unknown to non-resident stillwater anglers. Perhaps the other attractions, especially the many trophy brown trout rivers that are so famous are responsible. Make no mistake, Utah is not without its own premier stillwater fisheries. There are over 1,200 hike or pack-in lakes and streams situated in the Unita Mountains of northeastern Utah, many of which hold some big brook, rainbow and brown trout.

Most of the trophy hunters here spend their time on Strawberry Reservoir, a huge impoundment with a reputation for double-digit rainbows. The reservoir had some off years recently, but the rainbows are on a comeback. Cutthroats have been the focus of anglers the past few years, but with a lake as fertile as Strawberry, the big trout will be back soon.

Two other lakes that qualified for my Top 50 list are Panquitch Lake and Minorsville Reservoir. Otter Creek Reservoir, located down the road from Panquitch, still holds some very big trout, but is suffering from increased numbers of rough fish. There are countless others waiting to be discovered and as more and more stillwater fly fishermen reach out for solitude and pursuit of bigger trout, these hidden gems will no longer remain hidden.

When you reach that point where your arm won't take another cast, relax and enjoy the many scenic wonders Utah has to offer. The national parks here such as Bryce Canyon, Capitol Reef, Zion and Canyonlands offer spectacular views unparalleled on this earth. So, go stick a few and enjoy the country.

Mike Dehart
into a heavy
Minorsville
rainbow.

MINORSVILLE RESERVOIR
Utah

This is not a reservoir that is going to excite you at first glance, but wait until your line goes tight. Minorsville Reservoir is located in southwest Utah near its namesake. Most of the terrain is fairly flat, but with some soft rolling hills.

You won't find any 15 to 20 pound trout in this reservoir, but there is a healthy population of 2 to 6 pound rainbows with a few survivors reaching the 8 to 10 pound class. This is a rainbow fishery and they can put the weight on in a hurry when the reservoir is full of water. The main diet of these fish are chubs, leeches and crayfish. Anglers who fish the minnow or bugger type flies are going to be more successful than those who fish insect patterns because of the trout's food supply. Aquatic insects play a minor role except for midges that provide snack calories between meals. I have enjoyed good success using my olive Seal Bugger in size 8 during daylight hours and the black Seal Bugger before sunup and again the last hour before dark. My Callibaetis and A.P. nymphs provided some action, but most of the 'bows were less than 2 pounds. If you are lucky enough to find these fish chasing the chubs in shallow water, which happened one early morning for me, try a minnow imitation. I used both my Shiner and Chub minnow patterns and both took fish, but the bite was usually brief lasting only 20 to 30 minutes. Keep a keen eye on the shallow water shoreline areas. The water will explode with slashing boils and little fish leaping into the air to avoid the pursuit of these aggressive trout.

I found the best area in spring as well as the fall to be across the lake from the campground. The campground side is shallow, somewhat weedy and the home to thousands of chubs. The opposite side of the reservoir is rocky with drop-offs, the perfect place to ambush an unsuspecting meal.

Spring fishing, especially after ice out when it does freeze over, usually offers the best fishing of the year and can last well into June before the bite begins to slip. Trollers cover the lake at this time as well so you will have to practice patience to get along with each other. Sharing is nice, so show some respect for the other guy and give him some space.

The reservoir covers about 990 acres when full and gets an algae buildup during the warmer summer months. During the fall, as temperatures begin to cool, the algae will start to dissipate, improving oxygen levels, which allow these rainbows to begin feeding again with some regularity.

A variety of full sinking lines including the clear transparent, Uniform Sink II and intermediate are all good choices. Just match your line to the depth you plan on fishing. If fish are showing near the surface, the intermediate will be best but if no trout are showing, use the clear or uniform sinking lines and count down until you start getting hit. I didn't see much need for a floating line in the spring or fall, but if you want to fish dry, give it a try.

Leaders from 10 to 12 feet with tippets tapered to 3x to 4x will work just fine with any of these flies.

Minorsville Lake State Park is located on the lake with 29 sites and is 12 miles southwest of the town of Beaver on Highway 21. It gets cold here during the spring and again in the fall so go prepared.

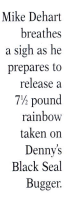

Mike Dehart breathes a sigh as he prepares to release a 7½ pound rainbow taken on Denny's Black Seal Bugger.

Minorsville Reservoir Summary

SEASON: Open year-round.

TROUT: Rainbow make up the fishery here with the average trout running 2 to 6 pounds and a few reaching 8 to 10 pounds.

LAKE SIZE: Minorsville Reservoir covers 990 acres when full.

RECOMMENDED FLY LINES: I found the clear transparent line that sinks about 2 inches per second the best line when I fished this lake in the spring of 1997. A full, fast sinker will also work well when fishing the drop-offs along the shore opposite the campgrounds.

RECOMMENDED FLY PATTERNS: My Seal Bugger in black and olive along with minnow imitations will work best on these big rainbows. Most insect imitations are going to take the smaller fish.

BEST TIMES: The reservoir fishes best after ice out into early June and again in the fall after the water cools.

NOTE: As of this writing, Utah state fish and game officials are considering poisoning Minorsville to eliminate the increasing chub population. The big trout are still there, but you may want to check the lake's status before planning a trip to Minorsville Reservoir.

Another nice 6 pound rainbow taken in May on an olive Seal Bugger and a type II clear transparent line. (Dave Nolte photo)

This 3½ pound Panquitch brown has some kin all fly fishermen would like to meet.

PANQUITCH LAKE
Utah

I first heard about Panguitch Lake from a Utah fly fishing guide who only spoke in whispers. "You are going to love this lake," he told me. "Big browns, rainbows and brookies," he said in an excited voice. He was right. When I fished here last, I didn't land any monsters, but I did manage a rainbow and a brown over 5 pounds and a few others from 2 to 3 pounds. Driving through the little town of Panguitch I inquired from a few locals about the size of the fish there. You should have seen their look when I asked the question, "How big are the trout in Panquitch?" One guy at a gas station replied, "Who told you about Panquitch," speaking sarcastically as if I shouldn't be there. But, others I talked with were very helpful and spoke of some big trout in the lake and cautioned, it wouldn't be easy. Seems these big trout are picky about eating and out of reach at times, but in the spring and fall each year rainbows up to 8 pounds and an occasional brown over 10 pounds are landed. Perhaps the biggest challenge is the large brookies that inhabit the lake. Fish to 4 pounds are not that uncommon, but the ones from 12 to 14 inches will keep you busy in between the bigger fish.

Panquitch is a good size lake, with 1,234 acres and offers shallow weedy shoreline areas, perfect for stillwater fly fishermen. This is a fertile lake with a habitat capable of supporting hatches of midges, Callibaetis mayflies, damselflies and dragonflies along with small forage fish.

Most of the rainbows in Panquitch average 15-18 inches, but this 4 pounder is a hint of what is lurking in the depths.

On my most recent trip here the wind blew and the hatches were nil, yet I still did well fishing my Stillwater Nymph in and around the weed beds. I also fished my Callibaetis Nymph and although I didn't land many fish with it that day, I did land one big brown on it. If I had stayed overnight, I would have used a streamer pattern to match the forage fish these big trout prey on in the early hours. Like all large trout, high protein food sources are necessary to sustain themselves so the forage fish are usually their main focus. Aquatic insects fill the gap between meals, but you need patterns to match both. Because of the vast shoal areas around the lake, the intermediate line, or when conditions call for it, the clear transparent line should be your first consideration. Floating lines are in order when fishing with dries or emerging chironomids. There are some deep areas in the lake where the bigger trout will obviously take shelter when the warm weather shows up. But remember, they only rest there, and rather than challenge a fish that isn't interested in eating while resting in deeper water, it is easier to catch trout in shallow water where they look for food.

Leaders and tippets are relative to existing conditions, but if in doubt, I'd recommend a 12 foot leader with a 4x tippet. That will get you by under most conditions. Wind is common here especially during the spring months so go prepared. If you fish here in late fall, overnight freezing temperatures are not that uncommon.

Panquitch Lake is located 19 miles southwest of the little town of Panquitch on State Route 143. There are two campgrounds on the lake, Panquitch Lake North with 49 sites and Panquitch Lake South with 19 sites. There are also several resorts located on the lake or you can find lodging and supplies in the town of Panquitch.

Panquitch Lake Summary

SEASON: Open year-round.

TROUT: Rainbow, brown and brook trout provide anglers with plenty of opportunity and variety. Rainbows are the most dominant trout with fish to 8 pounds possible. Browns will average 2 to 4 pounds with double-digit fish available. The brookies run 12 to 14 inches with fish to 5 pounds ready to test your skills.

LAKE SIZE: This is a good size lake with 1,234 acres to play in. There is plenty of weedy shoreline area to explore, perfect for stillwater anglers.

RECOMMENDED FLY LINES: The intermediate line is best for fishing the shallow weedy areas and I prefer the clear transparent type II full sinker to work the deeper areas.

RECOMMENDED FLY PATTERNS: I enjoyed the best action using my Stillwater Nymph in and around the weed beds. My Callibaetis Nymph or A.P. Emerger was a close second. This is a lake that is best fished with streamers or Seal Bugger type flies when the bigger trout work the shallows for forage fish.

BEST TIMES: Spring and fall are best for big trout, but the summer months can be excellent when hatches bring trout into shallow weedy areas to feed.

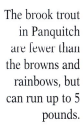

The brook trout in Panquitch are fewer than the browns and rainbows, but can run up to 5 pounds.

STRAWBERRY RESERVOIR
Utah

Dave Freel
and a
beautiful
evening
on Utah's
Strawberry
Reservoir.

Strawberry Reservoir is considered by many anglers to be Utah's best trophy trout fishery. That statement might summon the troops for a debate, but in years past there were few who would argue.

This is a huge lake, actually two separate reservoirs combined into one with over 8,400 surface acres when full. Intensely rich in aquatic vegetation, it has all the ingredients to grow big trout. In the simplest terms, this lake is a smorgasbord of aquatic insects and other food forms for anglers to imitate and for trout to feed on. Damselflies, scuds, leeches and forage fish are the staples of these trout with midges, dragonfly nymphs and Callibaetis mayflies available as well. Anglers using imitations to match these food sources will find the big fish receptive if matched with a solid presentation.

Strawberry's reputation over the years gained credibility because of its ability to grow trout to double-digit proportions. In the reservoir's early stages, rainbows were the beneficiaries of the rich habitat with fish from 8 to 12 pounds possible. But, fish and game officials stopped stocking rainbows several years ago replacing the species with surplus numbers of cutthroats. Today, the large rainbows have all but disappeared. However, recent plants of smaller rainbows give rise to the hope of a trophy rainbow fishery once again. In the meantime, cutthroats dominate the lake with trout to 6 or 7 pounds possible. A scattering of brook trout also remain a part of the fishery although fewer in number.

As with any lake there are some areas better than others and only time spent on the water will solve the mystery of figuring which spot is better. I've always found the weedbeds, shallow bays and the old streambed channels productive areas to explore both spring and fall.

The best time to fish the reservoir begins right after ice out and extends into late June. Summer temperatures will move the trout into deeper water with morning and late evening the prime times to find feeding fish. As soon as temperatures begin to cool, which usually begins in late September, trout return to the shallow bays and weed beds to feed. This is prime time when most of the bigger fish are vulnerable. Fish feed aggressively then and will continue on into November despite weather patterns that are often less than comfortable for anglers.

I have found my Stillwater Nymph or my olive Seal Bugger the best flies for Strawberry. That is especially true when the damselflies are showing around weedy areas and the shallows. If these fish are working the surface, I'll switch

A hefty 5 pound cutthroat from a quiet bay in Utah's Strawberry Reservoir.

to my tan Callibaetis Nymph or my A.P. Emerger and work them slowly with a hand-twist retrieve just below the surface on an intermediate line. The bigger trout will slam a well-presented streamer pattern in the fall. Normally, pattern selection isn't critical, but keep them sparse and in sizes 8 to 10.

To fish this big reservoir effectively, I use an intermediate line for the shallow areas and weed beds. I fish it down to 6 feet maximum. If the trout are deeper than that, I'll switch to a transparent type II sinking line and work the depths from 6 to 12 feet. If they are beyond that depth, they are not feeding and you will have to entice them. Leaders from 10 to 12 feet and tippets of 4x or 5x work fine.

Because of its size, it's a good idea to fish from a boat. Float tubes, pontoon boats or wading are all worthwhile alternatives but are limiting. Access is no problem with a dirt road circling the entire lake. Winds can pick up in a hurry and if severe enough, will keep anglers off the water sometimes for long periods. Because of its physical layout, some bays can be protected depending on the wind direction. Take the time to explore some of these coves. It could save the day when others give up.

There are numerous state campgrounds around the lake, but lodging, supplies and fishing information are some 30 miles away in Heber.

Big rainbows like Dave Freel's 8 pounder were once common in Strawberry, but now the reservoir yields more cutthroat.

Strawberry Reservoir Summary

SEASON: Open year-round.

TROUT: Cutthroats are the dominant trout with rainbows beginning to show in larger numbers. Brook trout are very much a minority, but show up occasionally. The cutthroats will run 16 to 20 inches with fish to 6 or 7 pounds possible. The rainbows, once the dominant species, run 12 to 14 inches with a few of the old hawgs still around.

LAKE SIZE: At 8,400 acres you have countless miles of shoreline shallows to explore with numerous coves and bays all accessible to wading or float tubes, but a boat is best.

RECOMMENDED FLY LINES: I feel the intermediate line will fish best in these shallow weedy areas, but a faster sinking line may be necessary to effectively fish depths of 6 to 12 feet.

RECOMMENDED FLY PATTERNS: I find my Stillwater, Callibaetis nymphs and A.P. Emerger best when fish are showing or when damsels are hatching. I'll use my olive Seal Bugger in size 8 to explore with early and late in the season.

BEST TIMES: This is a spring and fall lake for the bigger trout but summer can be good. Early morning and late evening are productive when hatches bring the trout to the surface.

Washington—Dry Falls Lake is one of many outstanding fly fishing lakes
in Eastern Washington.

Chapter 10

Washington

The best stillwater action in Washington is in the eastern half of the state. There are a few exceptions, but the Seep Lakes area is where the best fly fishing is concentrated. Cutthroats, rainbows and browns make up the action with most species averaging 3 to 6 pounds. Lenore is a well-known big cutthroat fishery while Lenice, Nunnully, Merry and Cherry are prime lakes for decent brown trout action. In my opinion the best rainbow fly fishing lake in the state is Chopaka Lake. In fact, many other anglers must share that opinion because it's never easy to find a camp spot here even in the off season.

Because of the climate, western Washington has limited trophy stillwater opportunities, however there are a few including Pass Lake near Bellingham as well as Coldwater Lake and Merrill Lake in Southwest Washington.

Sprinkled throughout the eastern region of the state are a number of quality lakes that didn't qualify for my Top 50 which are, nevertheless, very good stillwater lakes and reservoirs where you can test your favorite patterns. Don't hesitate to try O'Mack and Grimes lakes for big cutthroats if you have the time.

Because Eastern Washington is mostly dry and hot in the summer, concentrate your time in the spring and fall. There are exceptions, but you will find the angling opportunities best during these periods.

Regardless of where you fish in the spring, Eastern Washington is chironomid country and most stillwater anglers fish this insect pattern relentlessly. Of course, there are many other standard patterns that will work, but you will want to arm yourself with a variety of chironomids in different colors and sizes. Make no mistake, this little bug is big up here because of its numbers and the habitat that supports them.

CHOPAKA LAKE
Washington

Chopaka Lake is, in my opinion, Washington's most productive and best fly fishing lake.

Lying in a quiet little valley in the Okanogan Mountains, Chopaka is no longer an unknown fishery compared to other top stillwater lakes in Northeast Washington. This lake has to rank as one of the most pristine bodies of water in the entire state. With the Canadian border only a short distance away, Chopaka's appeal lies in its beauty as well as its outstanding fly fishing. Rainbow trout to 7 pounds are possible although few in number with the average fish 16 to 20 inches. Their aggressiveness and fighting ability is also a quality worthy of mention.

The lake is little more than a mile long with an abundance of shallow shoreline areas perfect for stillwater anglers to pursue cruising trout. It covers 149 acres with the deepest area about 35 feet deep. Hatches of Callibaetis mayflies are the main attraction and last from mid-June through early September. Damselflies and midges offer anglers some alternatives for these opportunistic feeding fish. An assortment of patterns matching these insects will work along with Leeches, Seal Buggers and Woolly Buggers in sizes 6 to 10. I've always found my Stillwater and Callibaetis nymphs in sizes 10 or 12 and 12 or 14, respectively, overwhelming favorites of these trout.

Early in the season (May and June) and again in the fall (September and October) my Stillwater Nymph and A.P. Emerger along with my olive Seal Bugger with burnt orange hackle are consistent and prolific producers of Chopaka trout. One particular trip, in May of 1997, I witnessed visiting anglers take up positions near the upper end and fish their favorite chironomid patterns. They used floating lines, long leaders and indicators. The method was deadly for those who figured out the right depth to set the indicator. It was consistent action from about 11 a.m. to around 2 p.m. Then, one by one they pulled anchor and departed. Why they didn't try other patterns I'll never know. From 2 p.m. to dark my party and I averaged better than 20 trout per rod, most of which ran 17 to 22 inches. Most were landed on my A.P. Emerger, Stillwater and Callibaetis nymphs although dry fly action can be outstanding in the shallows during evening hours all season.

Tubing seems to be the way to go and is the most popular method of fishing this lake, although some use boats. Wading from shore is very limited because of extensive tule cover after mid-June. Trout seem to be fairly well spread out throughout the lake, yet belly boaters appear to concentrate in the upper half of the lake.

When trout feed on or near the surface, a floating or intermediate line is best, but bring a medium sinking or uniform sink line for probing deeper water. In the summer, these trout have a tendency to drop into deeper water where it's cooler. Leaders of 10 to 15 feet are best, especially with floating lines when working the surface. Small tippets to match your pattern will bring more strikes, but exercise caution when setting the hook on these bigger fish.

Working the shallow shoreline areas offers excellent action for cruising fish. (Dave Nolte photo)

Chopaka's rainbows are strong and display a bad attitude when hooked.

Fishing seems to be consistent throughout the season with less pressure from anglers in late September and October. Night fishing is legal if you don't get enough during daylight hours. Camping is available and the spaces are well maintained, but difficult to find during the peak season. The road leading to the lake for the first mile is a bit steep and somewhat intimidating, but improves the closer you get.

Chopaka Lake Summary

SEASON: Open the last Saturday in April to October 31.

TROUT: Rainbow trout with the average 'bow running 16 to 20 inches and an occasional monster from 7 to 8 pounds.

LAKE SIZE: Chopaka is a long narrow lake with lots of shallow shoreline areas for stillwater anglers to explore. It holds 149 acres of water most of which support trout.

RECOMMENDED FLY LINES: A floating or intermediate line will fish this lake best. You can probe the depths with type II or type III full sinking lines, but the bite will be intermittent.

RECOMMENDED FLY PATTERNS: Most popular stillwater patterns work here. Midge patterns work well in the spring. I have had great success using my Stillwater Nymph, peacock Callibaetis Nymph, A.P. Emerger and for searching deeper water, my olive Seal Bugger when trout are not showing.

BEST TIMES: There isn't a bad time to fish this lake. You will find fewer fishermen in the fall, but the action is usually consistent unless weather patterns shut it down.

Dry Falls Lake, formed through glacial activity long ago, remains a big favorite among visiting fly fishermen.

DRY FALLS LAKE
Washington

Dry Falls Lake in Eastern Washington is not a lake with huge trout, but one of the better fly fishing lakes in the state. The lake got its name from a massive waterfall that poured basalt rock over its edge during the Ice Age, long before Mother Nature rearranged it again during the Glacial Age.

This is a unique lake ideal for fly fishermen who want to hone their skills while catching rainbow and brown trout in an unusual setting. With towering basalt cliffs surrounding the lake, there are approximately 100 acres of water to fish. The deepest part of the lake is about 60 feet with the average depth about 10 feet. Except for the rocky cliffs that extend into the water, the lake offers mostly a cattail shoreline.

Rainbows and browns are the object of fly fishermen with the average trout between 1 and 2 pounds and some reaching up to 5 pounds.

You will find midges on the water all season while mayflies seem to prefer a May emergence lasting into August as a rule. Damselflies begin to show when the weather warms usually by mid-June and will extend into September, although they're fewer in number.

An intermediate line, Stillwater nymph and a hand twist retrieve were too much for this 18 inch brown.

When to fish Dry Falls is directly related to the patterns that best match up with the most dominate insects available at the time. I usually fish this lake in June and again in September or early October. I've never been disappointed using my Stillwater Nymph in size 10 along with my A.P. Emerger or Callibaetis Nymph in tan or olive in sizes 10 to 12. These cover the damselflies and mayflies present in the lake at this time. You can expect early morning activity with trout feeding on emerging midges on or just below the lake's surface. That is when I like to use my Callibaetis Nymph and fish it in the top 6 inches of water. I use a hand-twist retrieve moving the fly very slowly in that zone. As the sun rises, trout will start dropping down so I switch to my Stillwater Nymph using the same retrieve, but fishing it down through the top 4 to 6 feet of water.

The intermediate line will fish these flies, or your personal favorites, in these zones better than any other line. You will need to use a 10 to 20 second count allowing your line and fly to sink until you find the depth the trout are holding.

I prefer using a 12 to 14 foot leader with 4x or 5x fluorocarbon tippet when I'm fishing this Grant County lake. I'm quite sure other lines and leader combinations will work, but this is what works for me. During the early part of the season, if the small nymphs don't match up because of lack of insect activity, I'll probe around using my Seal Bugger. That fly rarely fails to draw strikes. Bottomline: don't be afraid to experiment.

If there is a slow time on Dry Falls it would be during the warmer summer months when the surface temperature exceeds the trout's comfort zone and moves them into deeper water. But, you can count on early morning and late evening activity as the trout return to the surface to feed.

Since Dry Falls is managed as a quality fishery, flies and lures with single barbless hooks are required. If you need protein, you can keep one trout for the frying pan. The lake opens the last Saturday in April and closes November 30.

There is a campground and lodging at Sun Lakes State Park, or lodging in nearby Coulee City or Soap Lake. Dry Falls is located off Highway 17 with access through Sun Lakes State Park, which is about 4 miles from Coulee City on Highway 2.

A Dry Falls rainbow taking the afternoon off.

Dry Falls Lake Summary

SEASON: The lake opens the last Saturday in April and closes the end of November, but you might want to check the regulations in case these dates change.

TROUT: Rainbow and brown trout make up the fishery with the average trout running 15 to 18 inches and a few in the 5 to 6 pound range.

LAKE SIZE: Dry Falls covers about 100 acres with some shallow shorelines to pursue trout early and late in the day.

RECOMMENDED FLY LINES: I use the intermediate or type II clear transparent line to fish 6 to 12 feet down. You may need to experiment with faster sinking lines if you are not getting hit in the top 10 feet.

RECOMMENDED FLY PATTERNS: I've done my best using my Stillwater Nymph. My Callibaetis Nymph or A.P. Emerger are better choices when the trout are feeding on or near the surface. The Seal Bugger works fine when no fish are showing early or late in the day.

BEST TIMES: This lake will produce all season, but the better bite is spring and fall when the trout feed without being stressed by warmer water temperatures.

LENICE AND NUNNALLY LAKES
Washington

Lenice Lake is one of Washington's most challenging, yet rewarding stillwater fisheries.

These two lakes are twins separated by Merry Lake in between. Depending on how wet the winters, Lenice manages about 94 acres while Nunnally has approximately 120 acres. Both lakes are highly alkaline, infested with heavy weed growth, have prolific insect hatches and are rich in aquatic nutrients. Everything a fly fisherman could ask for is here, including big trout.

The Nunnally Lakes chain supports rainbows from 2 to 5 pounds with a few topping 7 pounds on occasion. There are browns here as well although fewer in number. Several years ago I witnessed a 12-year-old land a big male that was at least 10 pounds.

Lenice gets the nod for the biggest fish with browns reaching 6 to 8 pounds while the average rainbow is about the same as you'll find in Nunnally.

Both lakes support intense hatches of midges, caddisflies, mayflies, damselflies and dragonflies along with scuds and leeches. Hatches can be intense in the afternoons with mid-morning activity only so-so. Weather can curtail scheduled timetable emergence, but it's a nymphing and streamer fishery for the bigger fish anyway.

A nice brown from Lenice that found a Stillwater nymph a bit too enticing.

Favorite patterns include damselfly nymphs in sizes 10 or 12, Leeches, Seal Buggers, Woolly Buggers and Carey Specials in size 6 to 10 along with Zug Bugs and Hares Ear nymphs in sizes 12 to 16. A Midge Pupa and TDC nymphs in size 12 to 16 are fine choices when the trout are taking midges. My Stillwater Nymph has always done well for me here especially late May into mid-June.

Best times to fish these lakes are April through mid-June and late fall until it begins to freeze. Because of the searing summer heat, late July, August and into mid-September are usually slow. The season opens March 1 and closes the end of October.

Weather and the trout's natural spawning urge move the fish around, but when conditions are consistent, look for these trout to be actively feeding near the shoreline areas, weed beds and inlets.

A floating or intermediate line will work well on these lakes and if the trout move into deeper water, you may need to experiment with fast sinking lines. As with most lakes, let the conditions and feeding behavior of the trout dictate the pattern, line and retrieve speed.

Regulations call for artificials with single barbless hooks. There is a limit of one trout per day.

Merry Lake, located between Lenice and Nunnally, is an overlooked gem. Big browns abound here. Most anglers pass up Merry because it's smaller and shallower. I think they simply fail to recognize its potential. Merry is seldom easy, but can be extremely exciting. Early morning or late evening walking around the shallow weedy areas and fishing just off the bottom can be rewarding. Browns are structure huggers so use a line that keeps you in that area, especially when surface activity is non-existent.

To reach all three lakes take Smyrna Road east of Royal City, go west until you reach the parking area where signs are posted for Lenice, Merry and Nunnally lakes. If you hit the town of Beverly, you went too far. To get to these lakes you'll have to walk about a mile so be prepared to carry your gear. Because of the hike, fishing pressure can be on the light side. Most anglers use float tubes and a few hardy souls will portage a canoe into the lakes.

There is primitive camping in the parking area and lodging is available in Othello 25 miles away. Windy periods are common in the spring, but the hike is easy and the rewards are worth the effort.

Although browns are a minority in Nunnally, this 4 pounder is representative of the elder class in the lake. (Dave Nolte photo)

Lenice Lake Summary

SEASON: Opens March 1 and closes October 31

TROUT: Rainbow and brown with the rainbows running 2 to 5 pounds with fish to 6 to 7 pounds possible. The browns get bigger with some fish to 10 pounds.

LAKE SIZE: Lenice is about 94 acres with lots of shallow, weedy shoreline areas ideal for stillwater fly fishermen.

RECOMMENDED FLY LINES: You can explore the depths with fast sinking lines when these trout hug the bottoms to escape the warmer temperatures, but I still prefer the intermediate line for spring and fall fishing when these trout are feeding near the top or in the shallows.

RECOMMENDED FLY PATTERNS: I'm sure pattern selection isn't as critical as the depth you fish the fly. My Stillwater and Callibaetis nymphs are good when fish are showing near the surface or the Seal Bugger when they hold deeper.

BEST TIMES: This country gets warm quick, so I believe the best time (as with most high desert country lakes) is early and late in the season and early and late in the day if you fish the summer months.

This 5 pound rainbow was a springtime victim of a slow moving olive Seal Bugger fished in three feet of water.

Nunnally Lake Summary

SEASON: Opens March 1 and closes October 31.

TROUT: Rainbow and brown with the 'bows averaging 2 to 5 pounds and a few reaching 7 pounds. The browns average about the same, but fish to 8 to 9 pounds are possible.

LAKE SIZE: Nunnally is about 120 acres with plenty of weedy shoreline areas.

RECOMMENDED FLY LINES: The intermediate for shallow water fishing and a type II or type III when these trout go deep.

RECOMMENDED FLY PATTERNS: My Stillwater and Callibaetis nymphs are consistent when trout feed on or near the surface. The Seal Bugger or leech is good when probing the deeper water.

BEST TIMES: Same as Lenice. The bigger fish will be taken early and late in the season.

NOTE: All three lakes are currently suffering from an overpopulation of runt sunfish. The big trout are still there, but fewer in number. Funds are being collected (privately) to chemically treat all three lakes and return them to trophy status. Please check the condition of these lakes before planning a trip.

LAKE LENORE
Washington

Lake Lenore is an attraction for stillwater fly fishermen who are looking for a trout fix each spring.

High desert alkaline lakes are the focus of stillwater anglers in the Northeastern part of the state and Lake Lenore is no exception. With 1,600-plus surface acres, Lenore is the largest of the Sun Lakes Basin chain. The lake was once considered too alkaline to support trout until Lahontan cutthroats from Pyramid Lake in Nevada were introduced in 1977. The cutthroats flourished and Lenore is now considered one of the state's top trophy fisheries.

When the bite is on, you can expect numbers as well as size. Most of the cutthroats run 3 to 5 pounds, but 6 to 10 pound fish are landed almost daily during March and April and a few again in late fall. The lake is situated in barren, semi-arid country with little vegetation on the surrounding hills, but plenty below the lake's surface. Trees are non-existent, yet the lake is rich and fertile with an abundance of insect life. Chironomids, dragonfly and damselfly nymphs, along with leeches, are the most popular food sources for anglers to imitate.

The best patterns for me here have been my Seal Bugger, Midge Larva and Stillwater Nymph, all in olive. Experienced anglers prefer Chironomid Pupa in sizes 12 to 16, Carey Specials, Leeches, and Woolly Buggers in sizes 6 to 10, along with damselfly nymph patterns in sizes 10 or 12. Add to that lots of other local flavor and you begin to sense that pattern selection isn't all that critical. You're correct. More important is the depth you fish them.

Conversations with other anglers indicate most lines from floating to fast sinking including shooting heads and sink tips are all popular at Lenore. I've enjoyed good success using the intermediate and the clear transparent line that sinks about 2 inches per second. The line should match the pattern and retrieve you choose; it should also be fished at the depth these fish are holding. You will experience many situations where fish may be showing on the surface while others will be feeding on midges coming just off the bottom. The lake's diversity is one reason why it is so popular.

The best time to fish Lake Lenore is early spring beginning right after ice out if it has been frozen over and continuing through May, then again in the fall when the water cools. The bite is slow during the summer months beginning in late June and lasting through most of September. The trout seek cooler temperatures down deep and are unavailable to the anglers for the most part.

Cutthroats in general are not a difficult trout to catch, but when external forces are unsettled such as wind, barometric pressure, sudden change in temperature, north winds, etc., any trout can be temperamental.

Prime fly fishing areas are at both ends of the lake, but you will find fish cruising along the shoreline areas early and late in the day in both spring and fall.

Motors are prohibited on Lenore with most anglers using float tubes or pontoon boats. Wading can be very effective although you are more restricted in terms of reaching stratified fish in deeper water.

Sudden winds can be a problem so caution is advised. Fishing is catch and release only from March 1 to May 31. There are selective gear regulations for the remainder of the season until November 30. Also, there are posted closed areas at the north and south ends of the lake.

An average cutthroat from Lenore, but bigger ones are the object of most anglers.

Sun Lakes State Park is the primary camping spot a few miles north; however, there is lodging available in nearby Soap Lake and Ephrata. Other resorts in the area offer recreational vehicle hookups and camping as well. To reach Lenore from Ephrata drive on Highway 283 until you can turn on Highway 17. Follow 17 until you reach the lake. From Moses Lake, Highway 282 connects with Highway 17 which leads you directly to the lake.

Lake Lenore Summary

SEASON: Open year-round.

TROUT: Lahontan cutthroat trout averaging 3 to 5 pounds, but 6 to 10 pound fish are landed in the spring and fall.

LAKE SIZE: Lenore has 1,670 acres to explore with shallow shoreline areas where these cutthroats frequent spring and fall.

RECOMMENDED FLY LINES: Most lines work. I prefer the intermediate when fishing shallow, and a type II line for probing the depths from 8 to 12 feet. If you fish midges, a floater works well.

RECOMMENDED FLY PATTERNS: Leeches and Seal Buggers work well both spring and fall. Midge patterns fished with an indicator are popular among veteran anglers. I like my Midge Larva in black or olive, my Stillwater or Callibaetis nymphs in size 10 or 12 in olive.

BEST TIMES: This is an early spring or late fall lake when these big cutthroat move into shallow water. Nothing much goes on in the summer months when these fish are in deep water.

Chapter 10

Wyoming

Wyoming is a state rich in history, tradition and cowboy lore, but it is also blessed with some of the West's best stillwater fisheries. Much of the state is colored with herds of wild antelope, soft rolling hills, sagebrush and a dry, arid climate. But, the state is not without its high mountains and scenic wonders. It boasts such attractions as Yellowstone National Park, Teton National Park and the Wind River Mountain Range, home of the world record golden trout. My largest stillwater trout was also landed on the Wyoming side of Flaming Gorge Reservoir, a 23 pound brown in 1977.

The more I travel the state, the more impressed I am with the rich, fertile lakes and reservoirs found here. While fishing a high desert lake in 1993, a rancher rode up on horseback, tipped his hat and asked me how I was doing. I told him I landed a couple browns about 6 pounds and a rainbow about 4½ pounds. He wasn't impressed.

"Fella, if you want to catch some big trout, fish that lake up yonder," he said, pointing in the direction of a lake somewhere at the base of the Green River Mountains.

"How big?" I asked. "Big enough that you'll need more than that little stick you're waving around." He rode off and I've wondered about that lake ever since.

I'm quite sure as I travel the eastern part of the state I'm going to have to revise this book to accommodate some of the trophy lakes I haven't fished yet. Perhaps that will come in the year 2000. Perhaps someday I'll find the lake that cowboy was talking about.

One thing about Wyoming, you can't fish there without wind. If you do, it is the exception. I will promise you though, you won't be disappointed with the stillwater opportunities you will find in this state.

Wyoming's
Diamond Lake
is a popular
and consistently
good fly
fishing lake.

DIAMOND LAKE
Wyoming

There are many fly fishermen in Wyoming who consider Diamond Lake among the state's top stillwater fisheries. Whether that's true is a matter of debate, but Diamond certainly has all the ingredients to justify that ranking. It's shallow, fertile, alkaline, weedy and bug infested. Unfortunately, the days of double-digit lunkers are gone. Rainbows running 16 to 20 inches are the rule now with an occasional 5 to 6 pounder landed and trout to 8 pounds are possible. It is the constant action on fish to 20 inches that makes this lake so special. And if catching trout like that all day bores you, I recommend you fish elsewhere.

Diamond Lake covers 300 acres with 5 miles of shoreline. It lies in a shallow valley bordered by soft rolling hills on the northern side. Even when the lake is at its peak, it is rarely crowded. Ironically, I often see more license plates from Colorado vehicles than from the Cowboy State.

This is predominantly a rainbow fishery, however a brook or cutthroat trout is occasionally landed spicing the action. The lake is lined with extensive underwater vegetation allowing for dense aquatic insect growth, not to mention it is also great cover for the trout. Perhaps the reason more big fish are not landed is due to the extensive weed beds. It's easy to lose fish to the vast underwater gardens that cover the shoreline, but it's hard to resist fishing here. I think it's worth the gamble.

Because the lake is shallow, aquatic insects are plentiful with regular hatches occurring all season. Midges are almost always present, especially on cool overcast days. Callibaetis mayflies begin to hatch in late May and will thin by mid-September. Damsels begin around mid-June and last into September.

I last fished Diamond in June and again that fall a few years ago and I enjoyed exceptional action both times. My tan Callibaetis Nymph and A.P. Emerger were rarely refused if my presentation was good. Midges were plentiful both times and my Callibaetis Nymph did an excellent job of imitating adult midges. When adult Callibaetis were showing on the surface, I often used my A.P. Emerger and also fished it just below the surface. When these two insects began to fade from sight, I switched to my Stillwater Nymph to match emerging damsels and the "takes" just kept coming.

Lots of patience is in order here. The emphasis, as it should be, is on presentation. Having the right line, leader length, tippet size and the retrieve that matches the insect you are trying to imitate is the key to getting consistent action. For me, I used the intermediate line all day without changing because the bite was occurring just below the surface. I remember seeing several anglers using floating lines and indicators and doing well also.

The bigger trout lie deeper here so either countdown until you get hit or snag the weedy bottom. If the latter happens then back off some. I prefer using a 12 to 15 foot leader with 4x tippet, but you can experiment with that one depending on conditions.

The north end of the lake is the weedy area, but it is not the only area that will produce well. The entire lake is wadable and the little bays are popular foraging areas for hungry trout.

This rainbow is typical of the trout in Diamond Lake.

Diamond should fish well all season if lake levels don't fluctuate too much. Spring and fall should be best for bigger fish, but you have to spend more time locating them.

There is no closed season on Diamond. The limit, if you need it, is 2 fish per day. Any trout under 16 inches must be released unharmed and only flies and lures are legal. Is it any wonder why the numbers of big trout have dropped dramatically? Hmm???

At an elevation of 7,348 feet, it can get cold quickly and because you're fishing in Wyoming, wind can be an issue.

To get to Diamond, go about 6 miles past Arlington and take the Copper Cove Exit and turn left onto a county gravel road. Go about 4 miles then take the first left opposite King Reservoir #1 and it will lead you to the lake. There is unimproved camping with pit toilets and a 5 day maximum stay allowed.

Diamond Lake Summary

SEASON: Open year-round.

TROUT: Rainbow, brook and cutthroat trout make up the fishery. The rainbows are the most dominant species. A few cutthroat and brook trout are present, but their numbers are few. The average rainbow runs 16 to 20 inches and 'bows to 8 pounds are possible.

LAKE SIZE: When it's full, Diamond is about 300 acres with lots of shoreline areas to fish.

RECOMMENDED FLY LINES: I can't see any need for anything but floating or intermediate lines.

RECOMMENDED FLY PATTERNS: I've used a variety of my patterns here in various sizes and colors, all with good success. The A.P. Emerger, Callibaetis and Stillwater nymphs do plenty of damage. Pattern selection isn't as critical here, but the zone you fish is.

BEST TIMES: This lake fishes well all year, but look for the bigger fish to be more available in early spring or late fall.

East Newton Lake is a pristine, yet challenging stillwater fishery. (Dave Nolte photo)

EAST NEWTON LAKE
Wyoming

Located only 6 miles from the famous western town of Cody, East Newton has become one of my favorite lakes. It is as close to the ideal fishery as I can find; shallow, weedy, ideal water temperatures, scenic backdrop and small enough that it's easy to figure out after only a short period. With an abundance of aquatic insects and a healthy population of rainbow, brown and brook trout that average 18 to 23 inches with an occasional splake, you have it all. You can wade it, tube it, or fish it from a boat. All methods work. It is rarely moody and although the numbers of the bigger trophy fish from 8 to 10 pounds are far fewer now, there are still some around to keep you focused.

Tules, cattails and a variety of aquatic weeds provide the perfect habitat to support the food sources these trout eat and the cover they need to protect them from predators.

East Newton is not a big lake, roughly 125 acres with a maximum depth of 25 feet. That makes learning her secrets easier with fewer options to sort through.

Hosting trips into the Cody area twice a year, I always set time aside to fish this lake. Here, I have found some of the most beautiful trout in the western United States. While eating lunch at the water's edge in the fall of 1998, I watched a big male brook trout over 20 inches cruise by 10 feet in front of me.

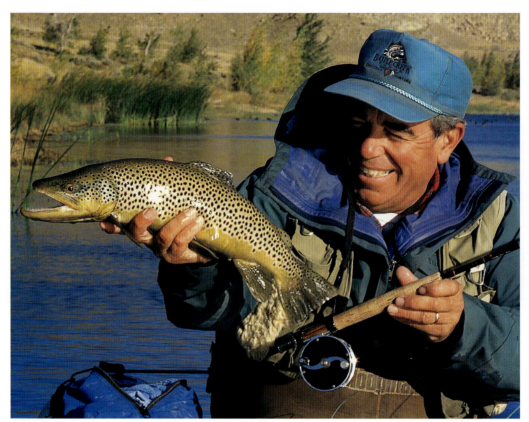

This 3 pound brown is about average for East Newton, but lurking in the depths are larger fish.

Spilling my Pepsi to get my rod, I could only watch as he slowly disappeared behind a stand of tules and into the depths. I know he is still there and I'll look for him again the next time I fish the lake.

At an elevation of 5,230 feet, East Newton offers stillwater anglers the perfect challenge. Although most of the local anglers choose to pursue these fish on foot, I feel you can cover more water fishing from a floating device. During the hot summer days, these trout will seek the deeper water well out of casting range of wading anglers.

My fishing partner, Dave Freel, prefers to use a type II clear transparent line when the trout are deep. He gives it a 30-second count and then imparts a slow retrieve with an olive Seal Bugger. He has taken browns to 27 inches and has left a few flies in the jaws of some that didn't want to see the inside of a net. Most of the time we use intermediate lines in and around the tules and weed beds.

Twenty fish days are pretty consistent with the majority of the trout running 2 to 4 pounds. I prefer to fish my Stillwater Nymph or olive Callibaetis Nymph in size 10 early and late in the day and my Midge Larvae, size 12 in olive or black, during mid-day hours both spring and fall. They are consistent for me and when the bite gets difficult or when trout are not showing, I'll use my olive Seal Bugger and go a bit deeper. I almost always start with leaders of 12 to 15 feet tapered down to 4x tippets. If trout are surface feeding, fish your fly in the top 6 inches and move it slowly. Strikes are often sudden and these trout can pull hard, so be prepared.

The most dominant insects here are damselflies and Callibaetis mayflies especially during the summer months. Trico mayflies and caddis hatches will spice the menu and offer you (and the trout) some additional options. Midges, scuds and leeches are always available to these fish so patterns imitating each should get their attention so long as you fish them at the depth where trout normally find them. For a brief period each spring, ants and hoppers make an appearance and the trout will respond accordingly. Fishing won't get any better than this if you time it right.

East Newton is managed for artificial flies and lures only with a one trout limit (20 inches or better). The lake is open all year, but the fishing begins unofficially at ice out. Spring and fall are the most consistent times, but weather can be on the wintry side and often unpredictable like most areas in the Rocky Mountains. Wind is also common this time of year, so dress accordingly.

To get to East Newton, take Highway 120 out of Cody and go northeast about 6 miles. Make a left turn at the rifle range and follow the road to the lake. There is also a sign on the right hand side of the road that will point you in the right direction.

Some of the most colorful rainbows I've ever caught were from East Newton.

East Newton Lake Summary

SEASON: Open year-round.

TROUT: Rainbow, brown and brook trout with an occasional splake (a cross between a brook and Mackinaw). The rainbow and brown will average 18 to 23 inches with some browns going 8 pounds and a few to 10. The brook trout are fewer in number with the average fish 14 to 16 inches. There are a few that reach 4 and 5 pounds.
LAKE SIZE: East Newton covers about 125 acres with shallow shorelines and weed beds around the entire lake.

RECOMMENDED FLY LINES: I use the intermediate for the shallow water fishing and the clear type II fast sinking line for probing the deeper water near the bottom.

RECOMMENDED FLY PATTERNS: My Seal Bugger in olive with a burnt orange hackle is very good both spring and fall. When trout work in shallow water and are surface or sub-surface feeding, I use my A.P. Emerger, Stillwater or Callibaetis nymphs in size 10 or 12 and fish them in the top 6 to 12 inches.

BEST TIMES: This one fishes well all year with the bigger fish showing early spring or fall.

Fontenelle
Reservoir is an
intimidating
body of water,
but the rewards
are worth the
effort.

FONTENELLE RESERVOIR
Wyoming

Fontenelle Reservoir is an intimidating body of water at first glance. In fact, it is just as intimidating at second glance. Fly fishing this lake is obviously a challenge, which is probably why it receives very little pressure from fly fishermen. The shoreline alone covers almost 40 miles and with over 8,800 acres of water to explore, it's hard not to get overwhelmed by its size. Disregard its massiveness and treat it as you would a smaller lake. Start at the two inlets and concentrate your efforts in shallow water where vegetation is prevalent. That's a safe and smart way to fish Fontenelle.

Fed primarily by the famous Green River, Fontenelle Reservoir grows large rainbows and browns. It also has a cutthroat population, but it is overshadowed by the carnivorous browns. The reservoir also boasts some big kokanee, which are the main target of fishermen in the summer who troll for these landlocked salmon. In the spring and fall, it's a completely different story.

Big rainbows are only part of the challenge that awaits stillwater fly fishermen on Fontenelle Reservoir.

Fontenelle boasts good numbers of rainbows from 2 to 4 pounds and there are 'bows to 10 pounds possible, but the lake's reputation comes from anglers who have landed the huge browns. In the past, fish over 20 pounds have been taken with 10 to 12 pound trout fairly common. Even the average brown will run 14 to 18 inches, respectable for most lakes, but not this one.

Fontenelle is a nutrient rich lake with a heavy algae bloom that lasts from mid-summer well into the fall. There is plenty of shallow water along the roadside shoreline and near the mouth where the Green River enters. Weed beds are present, but are not a major part of the trout's habitat. Midges are the dominant insect and are available all season with mayflies and caddisflies beginning in June and lasting into early October. While there's a myriad of insects available for the trout to feed on, this reservoir's trout grow to trophy size feeding on the kokanee fry. Because these trout survive and thrive on a high protein diet, I prefer using my Chub and Shiner Minnow patterns, along with my Seal Bugger, all in sizes 6 or 8. When fish are showing on the surface, I'll use my Stillwater Nymph in size 10.

It's important to ignore this lake's sprawl and to time your fishing trips when these trout are easier to locate. I find it more rewarding to challenge these big trout when they enter shallow water. That means fishing an intermediate or transparent type II full sinking line. If you need to get deeper, try faster sinking lines such as a type III or type IV. With those lines you'll be able to probe the deeper water using big bugger type flies and slow retrieves. When the algae gets thick, and it will start in June, try a 12 foot leader tapered to a 3x tippet.

As important as line, leader, pattern and retrieve are at this lake, timing and locating trout are even more important. If you miss on either, you can make a lot of casts with little or no chance of a trout seeing your fly. To improve the odds with these big fish, you have to pursue them when they're most vulnerable. With no closed season, you can stalk them early spring until late May, and again late in the season from mid-October until the lake freezes.

In the spring expect to find the rainbows in and around the channel where the Green River enters. There is also decent action near the mouth of Fontenelle Creek (the reservoir's other inlet) each spring. In the fall, these same places are most productive when the browns are preparing for their spawning run.

A daily limit of 6 trout is in effect, but only 1 can exceed 20 inches.

Highway 189 parallels the southern edge of the reservoir and there are several points that are accessible. A campground is located near the confluence of Fontenelle Creek and the lake. Lodging and supplies are available in Kemmerer about 50 miles southwest of the reservoir.

Fontenelle Reservoir Summary

SEASON: The lake is open year-round.

TROUT: Rainbow and brown trout with 'bows averaging 2 to 4 pounds with 10 pounders available. The browns are 2 to 5 pounds with 10 pound fish in fairly good numbers and some real trophies to 20 pounds possible.

LAKE SIZE: This one is big, over 8,800 acres with 40 miles of shoreline. There is a lot of shallow water along the edges, but the better shallow water is off the Green River inlet.

RECOMMENDED FLY LINES: The bigger fish are taken in shallow water, not out deep. I use the intermediate or clear type II line and fish down from 10 to 12 feet if there is no surface action.

RECOMMENDED FLY PATTERNS: This is minnow water so you can use leech, streamer patterns or Seal Buggers. When damsels show off the weedy areas, I use my Stillwater Nymph is size 10.

BEST TIMES: The best fishing is almost always spring and again in late fall when both rainbows and browns are in shallow water off the mouth of the Green River.

A nice Fontenelle brown, but much bigger fish are hidden in the vast regions of this huge reservoir.

The author with an average brown from Soda Lake taken on a Stillwater Nymph.

SODA LAKE
Wyoming

Lying at the base of the rugged Wind River Mountain Range, this gem of a trout lake is surrounded by unparalleled beauty. Snow is almost always present in the higher elevations and offers the camera buff the perfect backdrop showcasing a big brown trout in full spawning dress. Talk with some of the local residents and you will hear tales of 10 to 15 pound browns that helped form Soda Lake's reputation as one of the premier trophy lakes in the West dating back to the 30s and 40s.

The glory days are gone now, but Soda remains an outstanding fishery giving up a steady diet of 2 to 5 pound browns. There are bigger browns, some to 8 pounds, but their smaller brethren grossly outnumber them. However, browns aren't the only species of trout for you to stalk in this extremely rich lake. Brook trout averaging 14 to 16 inches share their space with an occasional lunker reaching 4 to 5 pounds.

For the fly fisherman who enjoys the pursuit of big browns and colorful brook trout, Soda Lake is perfect.

Aquatic weed beds cover the lake forming the perfect habitat for intense hatches of midges, mayflies and damselflies to imitate. Large numbers of Callibaetis mayflies and midges begin to show by late May, weather depending. Damsels typically make their appearance by mid-June and scuds are there for the taking throughout the year.

Roughly 312 acres, Soda has one small inlet channel where the browns show all year, but fishing pressure can be intense when spawning season nears. Because the entire shoreline is shallow, it's hard to find a bad fishing spot. That's especially true when the browns frequent the shallows foraging on the abundant bait fish. This is an excellent way to hook some of the lake's larger trout, especially if you get a wind blowing against the shoreline.

On a spring fishing trip a few years ago, my partner Dave Freel and I fished here and did quite well using intermediate lines and 12 foot leaders tapered to 4x tippet. Using different retrieves we both caught fish. I used a 15 second count down, a size 10 Stillwater Nymph and a slow, hand-twist retrieve while Dave used a 4-inch pull and pause retrieve. Both retrieves were met with aggressive takes from both the brook trout and browns. We didn't land any lunkers, although we both hooked a couple that would have qualified. Most of what we did land ran 2 to 3 pounds including a brown trout that went about 4 pounds. We also fished my olive Seal Bugger and Shiner Minnow patterns along the edges early and late in the evening and did quite well at times.

When the browns come to the surface to feed in Soda Lake, things get interesting.

Like so many of our better western lakes, Soda is a shallow lake. Instead of dredging the weedy bottoms with a fast sinking line, you can fish it effectively using the intermediate or the type II transparent line. Both lines will allow you to hold your fly above the weeds. Keep in mind that brook trout are basically bottom dwellers preferring some kind of structure for cover, so you will need to get down to them. Using the recommended lines, start with a 10 second countdown method and probe until you start getting strikes.

I've always found big brook trout a more difficult adversary than browns.

The season runs from May 10 to September 30. Outboard motors are not allowed from May 10 to May 31. There is a catch limit of 6 trout per day with one over 20 inches. These are the only restrictions on the lake and considering this lake's past and the fact that the majority of trophy fish are gone, the lack of restrictions and a generous limit is puzzling. Perhaps the Wyoming fish and game biologists need to take a hard look at their regulations and consider reducing the limits and eliminating the use of bait if they want to return the lake to its once storied past.

Because it is at 7,400 feet in elevation and in a valley, the lake is subject to high winds some of which could topple an out-house (there are several conveniently located around the lake). There is unimproved camping and most campers just find a level spot and claim residency for a spell. You will find motels in Pinedale, but you will find outstanding accommodations at Fort William, a step back into the past. This is a ranch with private waters and big trout. You can inquire by calling (307) 367-4670.

Soda Lake is located about 6 miles east of Pinedale on Highway 191. You have to travel a washboard, chuck-holed road that defies a pleasant description until the county fixes it. Pressure is seldom a problem here, so enjoy the fish, the scenery and the solitude.

Soda Lake Summary

SEASON: Opens May 10 and closes September 30.

TROUT: Brown and brook trout with the browns running 2 to 5 pounds and fish to 8 pounds possible. The brookies run 14 to 16 inches with a few 4 to 5 pounders available.

LAKE SIZE: Soda Lake has about 312 acres of water, but almost all its shorelines are shallow and weedy, perfect for the stillwater fly fisherman.

RECOMMENDED FLY LINES: This is an intermediate fly line lake or you can explore deeper with a faster sinking line. A type II will still keep you off the weedy bottom while moving parallel with slow moving retrieves.

RECOMMENDED FLY PATTERNS: I've had my best luck here using my Stillwater Nymph. I also enjoyed decent action with my Seal Bugger and Shiner Minnow patterns when I fished the shoreline areas, especially when the wind blows.

BEST TIMES: You can find action here all season, but the most consistent feeding is done in the spring into late June and throughout September.

YELLOWSTONE LAKE
Wyoming

Yellowstone Cutthroat may not be real big, but are very plentiful.

Yellowstone Lake is surrounded by Yellowstone National Park which requires no introduction. Folks traveling east or west usually plan to stop here to view its wildlife, scenic beauty and natural wonders. While this may be old stomping grounds to tourists, Yellowstone Lake is still an unknown to most fly fishermen.

This huge natural lake covers over 150 square miles and is home to the famous Yellowstone cutthroat trout. You won't find monsters in this lake unless they are the illegally introduced Mackinaw, but you will find all the 1 1/2 to 3 pound cutthroats you can handle. Because of its size, the lake can be intimidating. But there are actually lots of areas where these fish can be caught. Fish move in and out of the shallow bays and shoreline areas and seldom need to be challenged in water deeper than 10 feet. Some of the best fishing is in the south and southeast arms, but unless you enjoy hiking 15-plus miles, access is limited because powerboats are not permitted in either of these areas.

When you find fish, which isn't difficult, it's not uncommon to enjoy 30 to 40 fish days on Yellowstone. For those of you who want to sleep in, do it. The cutthroat are not picky as to when they look for calories. I've had success at

Yellowstone Cutthroats will average 16-20 inches and run 2-3 pounds.

all hours of the day. There are several prominent hatches for fly fishermen to consider and lots of aquatic insects to imitate. When the weather and water temperatures warm up to the comfortable level, which is usually by late June, Callibaetis mayflies and damselflies become active. Midges, leeches, scuds and dragonfly nymphs are also abundant all season. For fly fishermen, that means mid-June through October. If winter doesn't settle in early, this is an excellent time to imitate these food sources.

I prefer my Callibaetis and Stillwater nymphs in sizes 10 or 12 along with my olive Seal Bugger in sizes 8 or 10. Obviously, there are a lot of patterns that work on these fish. The key is to keep your fly at the depth the fish are feeding. Depending on the pattern fished you can use a floating, intermediate or type II or type III full sinking lines and all can be successful.

Fishing from shore, from a float tube or boat are all effective ways to pursue these trout. Rough water can occur quickly and at any time so use caution when boating this lake.

If you are a dry fly addict, the nearby Yellowstone River is outstanding dry fly water with excellent hatches coming off throughout the season. The mayfly and caddisfly hatches in the summer are legendary and prolific enough to keep you casting dry flies to rising trout all day.

Campgrounds and lodging are available throughout the area and along the lake. The season on the Yellowstone opens June 1 and closes November 30. Although Yellowstone Lake doesn't grow huge cutthroats, the solitude, wildlife viewing and the sheer numbers of fish will more than make up for the lack of size. Bring the family and enjoy.

Yellowstone Lake Summary

SEASON: Opens June 1 and closes November 30.

TROUT: Yellowstone cutthroat trout and Mackinaw. The cutts will average 2 to 3 pounds with fish to 5 pounds tops.

LAKE SIZE: This one is as close to intimidating as you will get, however it's easily accessible where the highway parallels the shoreline. With over 150 square miles to fish, you need only concentrate in the top 10 feet or the shorelines areas to be successful.

RECOMMENDED FLY LINES: If you fish in shallow water or just below the surface, use the floating line or the intermediate lines. To explore water below 10 feet, use a type II full sinking line.

RECOMMENDED FLY PATTERNS: I used my Stillwater and Callibaetis nymphs and my Seal Bugger and had no problem catching fish.

BEST TIMES: This one is easy. You pick it.

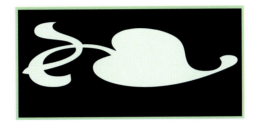

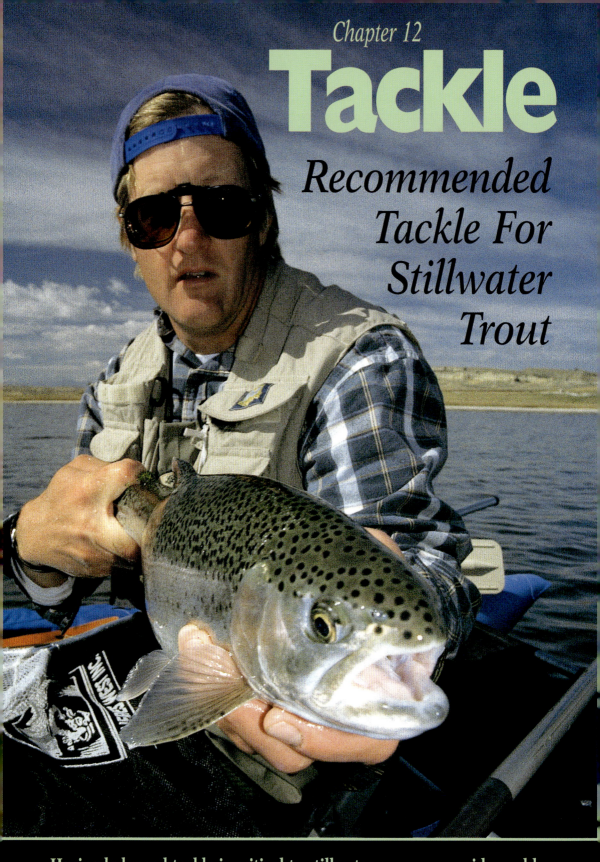

Chapter 12

Tackle

Recommended Tackle For Stillwater Trout

Having balanced tackle is critical to stillwater success as evidenced by
Tom Knudson's 6 pound rainbow. (Dave Nolte photo)

Choosing The Right Rod, Line, Leader and Tippet

Choosing the right fly line should be your first consideration when fishing lakes.

Stillwater tackle is not all that different from the gear we use for moving water, but the differences are significant enough to influence the outcome. Knowing which line, leader or tippet size to use is not only critical to a balanced presentation, but is often the difference between success and failure.

Fly fishing stillwater presents anglers with a variety of tackle options. Sorting out those options means matching the proper tackle to existing conditions both above and below the water. When done properly, it is the key to consistently landing big trout in lakes.

As stillwater anglers, the choice between the proper line, leader and tippet is just as critical to getting the "take" as our choice of fly pattern or retrieve. Whether we land the fish can be traced to rod choice along with our own fish playing skills.

My approach to stillwater angling is simple: I use a 9-foot, 5-6 weight graphite rod, two lines, five or six patterns, and basically three retrieves. With this system, I have fished lakes all over the country and have enjoyed relatively consistent success, not only with numbers of fish, but with size as well. Obviously, other skills such as being able to make the long cast, knowing

Making the right choices between line, fly and retrieve become reinforced when the strike occurs, not whether the fish is landed.

where to find trout, when to move, what depth to fish, along with how the external factors affect trout behavior are a big part of my system and the approach I take to fly fishing stillwater. All of the trout landed from the lakes and reservoirs listed in this book are the result of my stillwater system.

If you want to catch big trout, the first step is to fish where big trout live. Learn a system that will work, then spend time on the water improving it. Much of what you learn will be a result of putting in long hours on the water. As your commitment grows, your skills will improve. In time, hopefully, you will share that knowledge with a family member, a friend or a fellow angler. That, in my opinion, is an integral part of what fly fishing is all about.

RODS

When selecting a rod for stillwater angling, emphasis should be placed on length, weight and action. That means fishing with a rod that has some backbone. It should be long enough to provide the power for long casts, yet delicate enough to handle big trout on light tippets.

Long rods of 9 feet or more offer an advantage over shorter rods because of the leverage gained for making longer casts, a prerequisite for fly fishing lakes.

Part of the promise of fishing stillwater is the opportunity of hooking big trout. Challenging a big trout with fast action rods and light tippets is never a good idea. Using a soft tip rod may not guarantee break-offs won't occur, but it does reduce the risk of parting the leader by allowing the rod tip to absorb the bulk of an over-aggressive hook set. Fast action rods cast easier, are a better choice in strong winds, but are also less forgiving and too stiff in the tip section to compensate for the power and quickness of big trout.

All of the trout landed from the 50 lakes and reservoirs in this book were taken on my custom built, 9 foot, 5-6 graphite rod that incorporates a strong butt section, progressive action with a soft tip. With this rod, the stress of the set is no longer placed on the tippet, but on the tip section where it should be. Remember, rods should always match the tippets and flies first with regard to the size of the fish a secondary consideration.

FLY LINES

Because lakes and the habitat where trout live varies so much, a multitude of fly lines are necessary to explore all the options we encounter on stillwater. Floaters, sinkers, sink tips and shooting heads all have a function and a purpose. The key to success is knowing which line best matches the present conditions as well as the fly we are using at the time. That means choosing a line that will deliver your fly to the fish and hold it in the trout's feeding zone as long as possible.

The fly line dictates the depth and angle your fly must pass through the water during the retrieve. What good is the deadliest fly if trout don't see it? Think about it. There will be times when most lines, a wide assortment of flies and a variety of retrieves will all work. But, more often than not, trout, especially the wall hanger variety, are picky and selective about the times and places when they will look for food. The fly line is your key to presentation that makes it all happen. Trout feed near the surface and off the bottom in shallow water of any lake. By concentrating on these areas only two lines are necessary: the intermediate, which is the slowest full-sinking fly line available sinking about an inch per second, and a type II or the clear transparent fly line which sinks twice as fast. If you choose to fish with indicators or with adult fly patterns on the lake's surface, add a floater to your arsenal.

Since the greatest food bearing region of any lake is shallow water or near weed beds, no other line does a better job of fishing these areas than the slow sinking intermediate line. This fly line stays within the feeding zone for the greatest amount of time and with the proper retrieve, it presents your fly at the correct angle.

When conditions become difficult or when I need to fish depths greater than 6 feet, I'll switch to the clear transparent line. Making the switch allows me to probe the deeper water to about 10 or 12 feet. Over the years, I have found that if the trout are holding deeper than 12 feet, they are resting, not feeding. You can catch one now and then, but it is often unproductive fishing. With these two lines you can effectively match any condition found on lakes 95 percent of the time. Add in the right fly and proper retrieve and your potential for consistent success is unlimited. The intermediate and clear transparent lines are the only lines I used to fish all 50 of the lakes listed in this book.

Even the smallest trout in the lake won't take your fly if your leader and tippet are not balanced. (Dave Freel photo)

Note: In the fall of 1998, the Cortland line company introduced its clear transparent intermediate line which eliminates the need to switch lines from colored to transparent when clear, flat water occurs.

LEADERS AND TIPPETS

A well-designed leader is an integral part of a balanced system and is critical to a smooth presentation. Stillwater angling requires longer leaders and finer tippets than those we use for moving water. Most of my stillwater fly fishing is done with knotless leaders from 12 to 14 feet with 4x or 5x tippets. If conditions permit, I'll use 3x tippet depending on the size of the trout and the obstacles I encounter. When fishing with a high sun and clear, flat water, I may have to use 5x or 6x tippets and extend my leader from 15 to 18 feet to keep from spooking fish. With the new clear lines, you can reduce leader lengths to 9 or 10 feet since you essentially already have a 90 foot leader with the clear fly line.

What about tippet size and length? Is fluorocarbon better than monofilament? Controversy and debate are still very much a part of the pro's and con's when it comes to using fluorocarbon tippets and leaders. Is there an advantage in using fluorocarbon? I believe so. Fluorocarbon isn't as strong as some monofilament material, but I've found there are some conditions where fluorocarbon tippets will out-fish standard tippet material. I won't go as far to say this is the case all the time, but it has made a big difference at times. When nutrients in the water affect visibility, fluorocarbon tippets may not offer an advantage, however in clear water I seem to do better. Most of the time I'll use 3 foot fluorcarbon tippets.

However, the bottomline is this: Let water conditions and external factors dictate leader lengths and tippet size. Don't try to save a few pennies on leaders and tippet material. Buy the best. You won't be sorry.

Chapter 13

Flies

Suggestive
Patterns
For Lakes

Flies—When the Hexagenia mayfly is hatching, big trout pay attention.

Choosing The Right Fly

Trout fishermen are in love with trout flies. Perhaps it is this marriage that keeps us bonded to the sport. There is something magical about mesmerizing rows of neatly tied patterns combining fur and feathers that sparks conversations among anglers. Most of us, if we stay with the addiction long enough, will learn to tie our own. Imagine what a fitting salute it would be to your creative juices if your first trout were caught on a fly you tied.

More often than not we all tend to clone our flies to match the natural insect as close as possible. And why not? It's what we have been doing for years. The focus of fly fishing, as it has been since its inception, is to match the hatch, a term fly fishermen use to impersonate the living insect with one of man's creations. I would agree if it were the adult we are trying to copy. Selective feeding usually is associated with adult insects, those with wings and found on the water's surface. This kind of feeding behavior demands we match the natural's size, shape and color as closely as possible. But, when trout feed below the surface the focus must be on nymph patterns that are suggestive and display motion. Patterns that breathe, undulate and imitate the movements of the natural insect are going to be eaten more often than those that don't.

What good is the deadliest fly if trout don't see it?

Flies that attract trout should be suggestive, imitating a variety of food sources

Suggestiveness refers to a fly that imitates, not one, but several insects. We can accomplish this by the materials we use in tying the fly as well as altering our retrieves and line choices. This allows the angler to take advantage of the opportunistic feeding nature of stillwater trout, those whose size is measured in pounds not inches. Flies that move and breathe will catch the attention of trout and often trigger an aggressive response. When trout are selectively feeding on the surface, I do well fishing my nymphs at the beginning or toward the end of the hatch. At the height of the hatch, with so many naturals on the water, trout pretty much feed entirely on the adult. This is the time for a hatch-matching dry fly.

The flies represented in this chapter are the patterns I used to fish all of the lakes covered in this book. I found there were moments on many of these lakes and reservoirs that any number of standard patterns would have worked. But as important as fly selection is, I believe the depth and manner in which it is retrieved is far more important than what fly is plucked from your fly box. There will be times, however, when trout choose to feed selectively placing the emphasis on both pattern and presentation. During these moments, it is often not the pattern, but the size and color of the pattern that is the key.

Listed below are the patterns I use everyday on the lakes and reservoirs in the eleven Western states, Canada and New Zealand. They include a Seal Bugger, Leech, two minnow patterns and six nymphs. Because habitats vary, as do conditions on all lakes, I have included a variety of colors of these patterns that were used while fishing these top fifty lakes.

Pattern color is unimportant without light, but flies that breathe or display motion are quick to be eaten. (Dave Freel photo)

DENNY'S SEAL BUGGER #1

Hook: Size 6-10 4x long
Weight: 20 wraps of .020 wire at the head
Tail: Black marabou (somewhat sparse) with two strips of Flash-A-Bou
Hackle: Grizzly saddle hackle dyed burgundy wrapped 4 or 5 times
Body: Black seal fur or seal sub mixed with 1/4 red seal fur or seal sub
Rib: Copper wire
Head: Black

DENNY'S SEAL BUGGER #2

Hook: Size 6-10 4x long
Weight: 20 wraps of .020 wire at the head
Tail: Burgundy marabou with two strips of Flash-A-Bou
Hackle: Grizzly saddle hackle dyed burgundy
Body: Black seal fur or seal sub
Rib: Copper wire
Head: Black

DENNY'S SEAL BUGGER #3

Hook: Size 6-10 4x long
Weight: 20 wraps of .020 wire at the head.
Tail: Medium olive marabou with two strips of Flash-A-Bou
Hackle: Grizzly saddle hackle dyed burnt-orange
Body: Medium olive seal fur or seal sub
Rib: Copper wire
Head: Black

DENNY'S SEAL BUGGER #4

Hook:	Size 6-10 4x long
Weight:	20 wraps of .020 wire at the head
Tail:	Burnt-orange marabou with two strips of Flash-A-Bou
Hackle:	Grizzly saddle hackle dyed burnt-orange
Body:	Dark olive seal fur or seal sub
Rib:	Copper wire
Head:	Black

DENNY'S STILLWATER NYMPH #1

Hook:	Size 10-12 2x long
Weight:	6 wraps of .020 wire at the head
Tail:	Medium olive marabou
Hackle:	Grizzly saddle hackle dyed burnt-orange tied in tip first
Body:	Olive seal fur or seal sub, marabou or ostrich herl
Rib:	Copper wire
Wing Case:	Medium olive marabou tied down full length over entire body
Head:	Olive when unweighted and black when weighted

DENNY'S STILLWATER NYMPH #2

Hook:	Size 10-12 2x long
Weight:	6 wraps of .020 wire at the head
Tail:	Burnt-orange marabou
Hackle:	Grizzly saddle hackle dyed burnt-orange tied in tip first
Body:	Olive seal fur or seal sub, marabou or ostrich herl
Rib:	Copper wire
Wing Case:	Medium olive marabou tied down full length over entire body
Head:	Olive when unweighted and black when weighted

DENNY'S TAN CALLIBAETIS NYMPH

Hook: Size 10-14 2x long
Tail: Lemonside wood duck or dyed wood duck breast feather
Body: Hare's Ear
Rib: Copper wire
Hackle: Small grizzly hackle wrapped 3 times and tied up in tip first
Wing Case: Same as tail tied in full length of body
Head: Tan

DENNY'S PEACOCK CALLIBAETIS NYMPH

Hook: Size 10-14 2x long
Tail: Lemonside wood duck or dyed wood duck breast feather
Body: Peacock herl (3 strands)
Rib: Copper wire
Hackle: Grizzly saddle or cape dyed burnt-orange
Wing Case: Same as tail tied in full length of body
Head: Tan

DENNY'S OLIVE PEACOCK CALLIBAETIS NYMPH

Hook: Size 10-14 2x long
Tail: Mallard breast feather dyed dark olive
Body: Peacock herl (3 strands)
Rib: Copper wire
Hackle: Grizzly saddle or cape dyed burnt-orange
Wing Case: Same as tail tied in full length of body
Head: Olive

DENNY'S TAN A.P. EMERGER

Hook:	Size 10-14 2x long
Tail:	Wood duck or mallard breast feather dyed wood duck
Body:	Hare's ear
Rib:	Copper wire
Thorax:	Peacock herl (3 strands)
Wing Case:	Same as tail
Hackle:	Partridge (2 turns tied down over the thorax)
Head:	Tan

DENNY'S OLIVE A.P. EMERGER

Hook:	Size 10-14 2x long
Tail:	Wood duck breast feather dyed olive
Body:	Olive Hare's ear
Rib:	Copper wire
Thorax:	Peacock herl (3 strands)
Wing Case:	Same as tail
Hackle:	Partridge dyed olive (2 turns tied down over the thorax)
Head:	Olive

DENNY'S BLACK A.P. EMERGER

Hook:	Size 10-14 2x long
Tail:	Black moose mane or hackle barbules
Body:	Black seal or rabbit fur
Rib:	Copper wire
Thorax:	Peacock herl (3 strands)
Wing Case:	Same as tail
Hackle:	Partridge (2 turns tied down over thorax)
Head:	Black

DENNY'S BLACK DIAMOND

Hook:	Size 10-12 3x long
Weight:	10 turns of .012 wire at the head
Tail:	Black moose mane or hackle barbules
Body:	Black seal fur or seal sub
Rib:	Copper wire
Thorax:	Black seal fur or seal sub
Wing Case:	Turkey tail feather
Hackle:	One or two turns of black hackle tied down over thorax
Head:	Black

DENNY'S BLACK LEECH

Hook:	Size 6-10 3x long
Weight:	20 turns of .020 wire tied in at head
Tail:	Black marabou (sparse)
Body:	Black marabou
Wings:	3 wings of black marabou tapered back toward the head
Topping:	2- to 4 strands of pearlescent Flashabou
Head:	Black

This male brookie with its fall dress apparent was feeding in two feet of water when it took Denny's Olive Callibaetis Nymph.

DENNY'S BLACK MIDGE LARVA

Hook:	Size 10-12 3x long
Body:	5 strands of black marabou tied in tip first
Rib:	Copper wire
Thorax:	Black seal fur or seal sub
Head:	Black

DENNY'S OLIVE MIDGE LARVA

Hook:	Size 10-12 3x long
Body:	5 strands of olive marabou tied in tip first
Rib:	Copper wire
Thorax:	Olive seal fur or seal sub
Head:	Olive

DENNY'S BLOOD MIDGE LARVA

Hook:	Size 10-12 3x long
Body:	5 strands of red marabou tied in tip first
Rib:	Copper wire
Thorax:	Red seal fur or seal sub
Head:	Black

DENNY'S BLACK DRAGON

Hook:	Size 8-10 3x long
Weight:	10 turns of .020 wire tied in at head
Tail:	Pheasant rump dyed burnt-orange
Body:	Black seal fur or seal sub
Rib:	Copper wire
Hackle:	2 turns of pheasant rump dyed burnt-orange tied down over body
Head:	Black

DENNY'S SHINER MINNOW

Hook:	Size 6-10 4x long
Tail:	White marabou
Middle wing:	White marabou
Top Wing:	White marabou topped with gray marabou (add 4 strands of Flashabou over top)
Body:	Pearlescent shuck
Head:	Black

DENNY'S CHUB MINNOW

Hook:	Size 6-10 4x long
Tail:	White marabou
Middle wing:	White marabou
Top Wing:	White marabou topped with olive marabou (add 4 strands of Flashabou over top)
Body:	Pearlescent shuck
Head:	Black

Denny's Black Midge Larva and a slow pull and pause retrieve is a deadly combination for both trout and grayling.

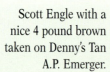

Scott Engle with a nice 4 pound brown taken on Denny's Tan A.P. Emerger.

The result of Denny's Black Seal Bugger and a good presentation.

This 9½ pound rainbow fell victim to the author's Tan Callibaetis Nymph. (Ray Beadle photo)

Chapter 14

Presentation

The Key for Success

Presentation—When it comes right down to it, presentation is the key to hooking trophy trout. (Dave Nolte photo)

The result of matching fly pattern and retrieve speed.

Matching Fly Line, Pattern and Retrieve

There are times when anglers who pursue trout in lakes find the bite difficult. It gets more frustrating when we see trout feeding, but can't get them to take. Most of us empty our fly box trying to figure it out. More often than not the problem isn't the fly, but what we are doing with it. From the trout's perspective, how the fly is presented, how it moves and acts, determines if he will eat it or not. When you decide to pursue trout in a stillwater environment with a fly rod, no other skill is more critical to success than your ability to make the fly act like the natural food source. In its simplest form, presentation is your ability to make trout eat. When we break it down, presentation is a three-part system that includes tackle selection, casting ability and retrieve styles.

The key to effective presentation is sorting through the options, then choosing the right tackle and techniques for the situations and conditions present. This means having balanced tackle that will deliver the fly to the fish, choosing a line that will suspend your fly for an extended period at the depth fish are feeding and retrieving it in a way that simulates the natural insect.

When I'm instructing students in schools or guiding clients on the water, I like to look at presentation in this way: Presentation begins with the line, which dictates the depth and angle your fly moves through the water. Although pattern selection is not part of presentation, it is difficult to think of one without the other. Pattern selection should be suggestive and match the food sources found in lakes. Once the cast places the fly on the water, the retrieve is what brings life to your fly. Of course, other factors must be considered, but this is where the emphasis on fly fishing stillwater should be placed. Catching trout then is a reflection of more time on the water.

Keep in mind you must always match pattern selection with the proper retrieve. If you use suggestive patterns, several retrieves can work with the same fly. To imitate different food sources, take into account the depth that the insect you are trying to imitate is found, then retrieve it in a manner that simulates the natural movements of that insect. Since we have already discussed the importance of line selection, let's look at the different retrieve styles that not only work on stillwater, but are the same ones I used on all the lakes while researching this book.

LONG SLOW PULL:

The long, slow pull simulates the idling indifference of many food sources found in stillwater. The length of the pull is not as critical as the speed you move your fly. When I'm fishing from a boat or wading, I like to use about a two foot pull with a definite pause between pulls. From a pontoon boat or float tube, it is difficult to move the fly two feet. I'll shorten my pulls to about a foot but maintain the timing for pauses and speed. Water conditions will dictate speed, but I find this retrieve deadly when I'm fishing a Seal Bugger, minnow imitations or leeches. This retrieve, coupled with those fly patterns, rarely triggers a smashing strike. The trout will usually suck in the fly which makes it crucial to keep a tight line.

Since trout feed in shallow water or near the surface, there is little margin for error when you invade their space.

Presentation is simply your ability to make trout eat your fly.

HAND-TWIST:

Having been around for a long time, the hand-twist retrieve is time tested. If you are going to fish lakes, you need this one. I have found the hand-twist to be the most consistent retrieve whenever I use flies size 10 or smaller, such as my A.P. Emerger, Stillwater and Callibaetis nymphs and my Midge Larva. The key to the retrieve is a slow, but consistent movement of the fly. Trout that are feeding on insects expect their prey to crawl or swim very slowly. All trout key on the insects movements and this retrieve does an excellent job of matching up. You may need to experiment a bit with the speed, but if ever in doubt, go slow.

SLOW 4 INCH PULL AND PAUSE:

This retrieve moves the fly much like the hand-twist only the pause allows the fly to rest and dip slightly. I prefer it when fishing my Dragon and Diamond nymphs. Since trout prefer to inhale their food, not strike at it, slow moving patterns are going to be eaten more often that those that move too quickly. This retrieve, like the hand-twist, is best with small flies size 10 to 16. Depending on the zone I'm fishing, I'll often use long pauses between pulls with this retrieve when the bite is off or inconsistent. Don't be afraid to explore some with this one, but the speed of the pull should remain slow.

RAPID 2 INCH PULL:

This is my go-to retrieve when trout display an aggressive attitude towards their prey or if the standard retrieves are not working. You can use this one with just about any fly, but it should always be used in the top 2 to 3 feet of water and when water temperatures are above 55 degrees. Keep in mind trout are not as aggressive when temperatures are on the cold side. The retrieve is a constant rapid movement in little pulls that almost always draws a vicious response. It can be used effectively in shallow water or around weed beds and often gives the trout the appearance of something trying to escape. Light tippets will not hold up well with this retrieve. You will remember that after a big fish breaks you off.

STREAMER RETRIEVE:

This is the retrieve I use when fishing my minnow imitations. It is very similar to the long slow pull I use with my Seal Buggers. The difference is this retrieve should be moved a bit faster to imitate a darting minnow. Because there is a wide variety of forage fish for trout to eat in lakes, you need to modify the distance and speed to match whatever the minnows are in the lake you are fishing. Remember to take into account the conditions present. This will leave a lot of options to choose from. Most of the time I prefer a relatively long semi-fast jerky pull with short pauses during the warm summer months. During the spring, or when water temperatures are below 50 degrees, I'll slow down using shorter and even slower pulls.

Remember, minnows use the shorelines for cover and trout only frequent these areas to feed when they have some kind of cover such as semi-darkness, ripple or algae. If you are getting short hits, the problem can be usually traced to hook size or speed of the retrieve.

There are two keys to presentation that need to be adhered to regardless of the retrieve you use. Always place the tip of your rod into the water, which keeps your line tight. You will miss fewer strikes and suffer fewer break-offs that way. Secondly, always remove the slack from your cast immediately after you line hits the water. Pull in the slack until you see your fly move. Many strikes occur as the fly is dropping without any movement on your part. Remember, you cannot set the hook, play a fish or retrieve your fly effectively when slack is part of your presentation.

Summary

Please keep in mind that being successful on stillwater is really a matter of choosing the right fly line to effectively fish the zone fish are holding in, selecting a fly that is representative of the food sources present and using a retrieve that emulates the movements of the natural.

That folks, is what I try to accomplish on every lake I fish. Obviously, there is more to this game than line, pattern and retrieve, but this is where the focus belongs. Keep in mind some lakes fish better early or late in the day and especially early or late in the season. This is critical when you are a predator of big trout. If your pursuit is for the trophy members that inhabit any of these fifty lakes, plan your trip carefully. It is always a good idea not to skimp on quality leaders and tippets. Use fluorocarbon when necessary and pay close attention to prevailing conditions. They are always an indicator of trout behavior.

Some of the fifty lakes and reservoirs listed in this book have gone through recent changes in an attempt to improve the fishery for the coming years. If the habitat has not been adversely affected, the fishery will return to past status in a short period. Hopefully you will find these lakes as productive as I have.

Now, go fish and enjoy!

Hank Bauma of the H Lazy 6 Ranch in Montana shows what a sound presentation can do.

Conclusion

When tomorrow comes, I'll probably be fishing one of my favorite lakes here in the West. Perhaps, it may be one I'm not so familiar with. A big part of the enjoyment I get from fly fishing stillwaters is eliminating some of the mysteries of a new fishery. I'm quite sure as I travel across the Western states there will be many new lakes and reservoirs waiting to replace some of these I've listed in my Top 50. It is a venture filled with promise and one I'm already looking forward to.

When we reflect upon our environment and the habitat our trout must endure to survive, we see lobbying, speculation, crisis and litigation regarding our favorite pastime. But, we are making strides and enjoying some progress as well.

Michael Furtman, author of "Trout Country" wrote in his dedication, "To those who cannot live without wild trout, and to those who fight to ensure we never face that prospect." To those organizations and people who work relentlessly without recognition in this arena, I salute you. Your efforts are giving and will continue to give direction for all who enjoy the sport of fly fishing in the years ahead.

Please keep in mind the future of our sport lies with the youth of today. I trust each of you will take the opportunity to teach and encourage your friends, sons and daughters the ethics of catch and release. And, let's not forget our on-the-water etiquette. Give your fellow angler his space and extend a helping hand to all who ask. This will be our legacy to those who follow, as well as those who push the buttons of progress tomorrow.

Here is wishing you the reality of the trout you dream of, but have yet to land.